CW00923001

Tank

Tank

The 10 War Machines That Changed the World and the Remarkable Men Behind Them

MARK URBAN

VIKING

UK | USA | Canada | Ireland | Australia
India | New Zealand | South Africa

Viking is part of the Penguin Random House group of companies
whose addresses can be found at global.penguinrandomhouse.com

Penguin Random House UK,
One Embassy Gardens, 8 Viaduct Gardens, London SW11 7BW

penguin.co.uk

First published 2025
001

Set in 12/14.75pt Bembo Book MT Pro
Typeset by Jouve (UK), Milton Keynes
Printed and bound in Great Britain by Clays Ltd, Elcograf S.p.A.

The authorized representative in the EEA is Penguin Random House Ireland,
Morrison Chambers, 32 Nassau Street, Dublin D02 YH68

A CIP catalogue record for this book is available from the British Library

HARDBACK ISBN: 978–0–241–74150–4
TRADE PAPERBACK ISBN: 978–0–241–74149–8

Penguin Random House is committed to a sustainable future
for our business, our readers and our planet. This book is made from
Forest Stewardship Council® certified paper.

To those who Fearnaught

Contents

List of Illustrations

Plate Sections

In Text

CREDITS

Plate Sections

Images 1, 2, 4, 5, 6, 7, 9, 10, 14, 15, 16, 17, 18, 19, 22, 23, 24, 25, 26, 27 and 36 reproduced courtesy of The Tank Museum. Images 3, 11, 13, 20, 21, 32, 34, 39 reproduced courtesy of Alamy. Images 12, 33, 35, 37 reproduced courtesy of Getty. The remaining images are in the public domain or are the author's own.

In Text

Images of the Renault FT and T-34 reproduced courtesy of Bridgeman Images. All other integrated images reproduced courtesy of Alamy.

Preamble: Dreams of Victory

Early on the 8th of June 2023, the Ukrainian columns moved through the darkness, down to their assault positions. Their mission was to breach lanes in the formidable Russian defences, the so-called Surovikin Line.

There were ploughs, turretless tanks that would lift the mines out of the way. Close by, Bradley Fighting Vehicles would cover them, while carrying forward assault infantry. Most impressively, there were Leopard 2s also, 62-tonne behemoths, more advanced than anything the enemy might put in their way.

Months earlier Ukraine's leader, Volodymyr Zelenskyy, had deployed his considerable skills of persuasion to obtain this precious armour. It was key, he believed, to breaking the Russian occupation. If his freshly trained, Nato-equipped brigades could pierce the Surovikin Line, then who knew, some of the optimists around him speculated, maybe they could push south to the Sea of Azov, slicing the enemy army in two?

That June morning they were trying to do just that. It was a set-piece attack spearheaded by armour in an area called the Orikhiv axis south-east of the Ukrainian-held city of Zaporizhzhia.

'The key thing now is speed and volume,' Zelenskyy had said in January 2023 when, after months of pressure, Germany had finally agreed to greenlight the Leopards. 'We must form a tank fist, a fist of freedom whose blows will not let tyranny stand up again.'

This objective – obtaining modern tanks, particularly the Leopards – became so important to him that Ukraine accepted British Challenger 2s that it didn't really want as a means of putting pressure on Germany to follow suit. And the United States agreed to send dozens of its own M1 Abrams in a parallel thrust to outmanoeuvre and shame the chancellor in Berlin.

The precious Leopards, as well as dozens of American-made Bradleys, had been given to a newly formed brigade, the 47th. 'We were told that either we will go down in history, or we will be forgotten,' Yaroslav, one of the unit's assault group commanders, told a Ukrainian newspaper, 'because we are here at the tip of the spear, we were to be sent into one of the toughest fights that this war could possibly offer.'

As the tanks moved down into the minefields one thing after another went wrong. The Russians started bringing artillery fire down on pre-registered points, the approach routes the Ukrainian columns were using. The blasts immobilized a couple of vehicles. While trying to move past them, another explosion: a Bradley had run over a mine outside the cleared lane, losing a track. Inside, the crews had survived but one by one their vehicles were being stopped.

Video taken after first light by a Russian drone revealed a mine plough tank, four Bradleys and a Leopard, all immobilized. Colonel General Alexander Romanchuk, commanding the Russian army in the sector, told a reporter that first his artillery had neutralized the Ukrainian long-range guns.

Once daylight showed the attackers' predicament, 'the crews of anti-tank missile systems, special forces, and combined arms units at the forefront, joined in the fire engagement of the enemy', said Romanchuk. In video taken by a Ukrainian Bradley crew sent in to rescue men from the immobilized vehicles, one of the Russian anti-tank missiles strikes the frontal armour, a blast the vehicle survived.

'The enemy was forced to stop,' the Russian general said, 'and having suffered losses, could not overcome the minefields.' Day by day, segments of video emerged, giving a wider context. Not far away, another four knocked-out Bradleys were visible, elsewhere two more mine ploughs, and another Leopard apparently damaged.

For observers far away collating these accounts it was becoming clearer and clearer that the 47th Brigade's assault, upon which months of preparation and the very best equipment Nato could muster were lavished, had fallen apart. The final straw came when a Russian Telegram social media channel posted video of their troops crawling over the knocked-out Western vehicles, looting them.

Ukraine had committed its precious armoured reserve but it had failed to fulfil its promise. In the aftermath there were recriminations over whether the Americans had pressured them into mounting a counter-offensive under circumstances where success was highly unlikely.

Zelenskyy had himself fallen into a tank trap. In that sense he was following in the footsteps of Winston Churchill no less, believing that tanks provided the solution to breaking a bloody stalemate on the battlefield, just as the British leader had in 1915 when giving his backing to some of the early experiments with armoured vehicles. Zelenskyy had bought the idea that these huge machines, in which the crews are protected from shot and shell, riding on tracks across muddy land impassable to normal vehicles, could dramatically change the battlefield calculus.

In 2025 the tank, an invention more than a century old, embodies technology dramatically different to the models first thrown into battle in 1916; from secure radio to night vision devices, enormously powerful engines able to drive them forward at ten times the speed of those early machines, to armour more than 40 times as thick. And yet there are constants.

Used correctly at the right time and place, tanks can still achieve breakthroughs. Major General Kathryn Toohey, tasked by the Australian army to examine whether these expensive monsters were still worth buying in the 2020s, concluded: 'tanks are like dinner jackets. You don't need them very often, but when you do, nothing else will do.'

Those preconditions for the successful use of tanks have remained remarkably similar throughout more than a century of their existence. They involve slick teamwork between all players on or over the battlefield: armour, artillery, infantry, aircraft and engineers. In that contest of rock, paper, scissors, there will be moments where the tank's blend of firepower, mobility and protection will prevail, and there will be others when it does not.

Writing in 1933, General George S. Patton, the first man enrolled in the United States Tank Corps, had seen something that perhaps Zelenskyy could not: 'History is replete with military implements each in its day heralded as the last word – the key to victory – yet each in its turn subsiding to its useful but inconspicuous niche.'

The Surovikin Line, protecting the logistics jugular of the Russian occupation army, represented a carefully constructed defence in which all arms played their part, which would require, equally, the effective use of every element of the Ukrainian military in harmony. Sending some modern tanks or Bradleys was never in itself going to do it.

'The breakthrough battle imposes some pretty tough demands on the tanks. Success is probably only attainable when the entire defensive system can be brought under attack at more or less the same time.' Those words, penned by German General Heinz Guderian in 1938, speak to what was lacking in the Orikhiv axis in June 2023 – sufficient airpower and artillery to blanket the Russians defending these obstacles with suppressive fire.

The sight of Ukrainian armour, languishing in the fields with blown-off tracks, brings to mind other advice given by Guderian in the same passage: 'Mines are . . . an enemy of an extremely dangerous order, and we must clear them at least in part before the armoured assault proper can break into the infantry combat zone.'

If the principles have remained similar, perhaps surprisingly so, across the last century, their application, as well as the development of armoured vehicles, has evolved relentlessly. At times, for example in the Second World War, the advances in gun or engine power were so rapid that a new machine might become obsolescent within a year of being fielded in battle. The advent of the tank triggered intense competition between industrial economies, a constant striving for gains that might make all the difference.

In the popular imagination, the tank became a terrifying beast of war. But its story, like that of any machine, is an amoral one. The rumbling leviathan became a potent symbol of fear or oppression as well as one of liberation and hope. It was forged as a tool for Soviet Communism as well as German Nazism, while at the same time symbolizing the 'arsenal of democracy'.

For decades my own life has been interwoven with this narrative, starting with my enlistment in 1979 at the age of 18 in the Royal Tank Regiment. I didn't serve for long, and it was in my subsequent career as a journalist, covering wars around the world, that I saw what machines expertly designed and crewed could achieve, contrasted with the price of failure. A youthful enthusiasm for armoured vehicles is quickly tempered by the sight, sound and smells of a knocked-out tank blazing on the battlefield, its crew having perished inside.

This is not a simple narrative either – of a linear rise or fall. Scepticism surrounded the birth of the tank, and it remains today. Predictions of its demise have punctuated its story. As we will see, the curve of armour's battlefield utility rises and falls as technology and other factors evolve.

Such is the complexity of this equation of success or failure that a comprehensive history of the tank, taking in all the relevant factors from metallurgy to morale, would require hundreds of thousands of words, filling many volumes. So instead in this book I have picked ten salient examples, outstanding designs that mark key moments in this wider narrative.

It is the story of an unceasing arms race. And it is also about the people who made the tanks, and how the ideas on what tanks are for, how exactly they fit with the other players on the battlefield, combined with the unending search for a trump card in the quest for victory.

Mark IV

Entered service: 1917
Number produced: 1,220
Weight: 28 tons (28.4 tonnes)
Crew: 8
Main armament: two 6-pounder (57mm) guns
Cost: £5,000 (£291,000 at 2024 prices)

Mark IV

Dawn did not bring its usual relief for the sentries on the Hindenburg Line that murky November morning. Far from it. The soldiers of Infantry Regiment 84 were about to fall victim to one of the most awful inventions of the industrial age.

'All of a sudden the British brought down appalling drum fire on our positions, forcing us to take cover in the nearest dugout,' Sergeant Schwarz, occupying K1, one of the forward trenches, recorded. A young British officer, preparing to cross no man's land, looked ahead: 'a solitary gun went and then the whole world lit up, it was Crystal Palace on the German lines.'

Crouching in their bomb shelters while British shells dropped around their positions, the German soldiers were confident. This stretch of front, near the French town of Cambrai, was so quiet that the General Staff sent units there to recover from hard fighting elsewhere. What's more, the officers of the 84th Regiment didn't much rate the British army and knew that around half-way between the two lines was one of the most imposing barbed wire obstacles on the entire Western Front.

Ten minutes into the bombardment the Germans had settled down, expecting a long stay in their bomb-proof shelters. But Lieutenant Norman Dillon had already covered the distance between the British start line and the enemy obstacle: 'I'd never seen such a depth of barbed wire . . . so dense with wire that you could barely poke a broom handle through it.'

From a distance, it had looked like a black wall, but close up the British could see the wire was around 15 feet deep, and anything up to 9 feet high. Any idea of clearing this entanglement, secured to the ground by metal pickets, with less than a week's artillery bombardment seemed inconceivable to the German officers.

Around 15 minutes after the shellfire had started, the British gunners began to shift the barrage deeper into the German defence, allowing the soldiers in the K1 trench to venture out of their dugouts. Sergeant Schwarz went back to the firing step of his trench. 'The sight that greeted us was completely unexpected. About twenty to thirty tanks were bearing down on us, and they were only about fifty to sixty metres away.'

Further along the K1 line, Lieutenant Saucke saw a panicked soldier jump into their trench. There was still a good deal of smoke in the air, and Saucke was getting his bearings after the shelling. 'A wounded man from the 7th Company approached us from the right,' the lieutenant recorded, 'and gasped a few heavily charged words, "the British have got tanks!" A cold shiver ran down my spine: the effect of this on the morale of my men was plain to see. They who had been pouring scorn on the British . . . suddenly looked disconcerted.'

Lieutenant Dillon had watched the vehicles of his battalion crushing down the formidable barbed wire obstacle in moments, and now beheld the spectacle of scores of Mark IV tanks nearing the German lines. Unlike much of the Western Front, drenched in blood and suffering during four years of fighting, the ground in the Cambrai sector had not been pulverized by months of shellfire. Dillon, who as the company reconnaissance officer was walking into action behind one of the leading tanks, recorded that progress was 'splendid, it hadn't been fought over much, the going was very good, and one couldn't have asked for better conditions'.

The evening before, the commander of the tank attack, Brigadier Hugh Elles, had issued a special order to his men. The advocates of tank warfare had experienced more than a year of disappointments and much scorn from headquarters. 'Tomorrow,' Elles wrote, 'the Tank Corps will have the chance for which it has been waiting for many months, to operate on good going in the van of the battle.' Lo and behold, on the 20th of November 1917, his vision of a successful mass attack using these new machines of war had come to pass. After earlier assaults featuring a dozen or two machines here and there, nine battalions of tanks, 378 fighting vehicles and 98 others (mainly used for carrying supplies forward), had been brought together for the Cambrai offensive.

Watching this force in full flood across the French plain, the Germans struggled to describe it. Some said the slow onslaught of Mark IVs looked like moving buildings. One compared them to huge blackboards; another, a soldier of the 84th Regiment, reported to his officer: '*Herr Leutnant*, something four-cornered is coming!'

Facing the Mark IVs in his trench, Lieutenant Saucke of that same regiment had started to direct machine gun fire onto the approaching behemoths. 'We saw, looming out of the swirling fog, a dreadful colossus heading straight for us . . . As I pulled myself up to look over the parapet, I could see a whole chain of these steel monsters advancing towards us.' Although some of the German riflemen and machine gunners had been issued with armour-piercing bullets, nothing seemed to check their progress.

One of the British gunners in tank F23 heard the German rounds splattering on the outside, making a 'noise like hail'. Another gunner, Private Jason Addy, later recounted the advance: 'the bus was spattered repeatedly with left and right sweeps of machine gun fire. The whole panorama now was like a set piece, with thousands of fountains of fire spouting from the earth.'

The lead tanks had reached the K1 line. Following the earlier use of British tanks, which the Germans knew had been built to cross trenches up to 9 feet wide, many of the forward trenches, including K1, had been widened to 12 feet.

With the armour almost upon them, and all their machine gun ammunition fired off, Lieutenant Saucke and his men abandoned their section of the forward obstacle, scurrying back through gunfire to the safety of K2, the second line of defence. The tanks, he was confident, would not get past the 12-foot gap posed by K1.

Peering over the parapet the young lieutenant watched one of the tanks, gingerly edging forward, tilting into the abandoned trench. It was only then that he and others understood the purpose of the huge bundles of brushwood, called fascines by the British, that had been secured to the roofs of the Mark IVs. Steel hawsers holding it in place were released, and the fascine tumbled into the K1 trench.

Watching it pitch forward, Saucke recalled, 'I assumed the tank had been hit or at least damaged, and could scarcely believe my eyes when it continued onwards and its outline became more clearly defined. There could be no doubt, it had crossed the trench and was pressing on towards us.' The leading tanks had used their fascines to bridge the 'uncrossable' K1 gap, and were heading for the second line.

Each platoon of four tanks now followed a well-rehearsed drill. The lead vehicle, minus its fascine, stopped a short way forward of the K1 trench, training its weapons on the machine guns spitting fire at them from K2. The second and third tanks, crossing the fascine laid by the first, turned left and right, driving along the K1 trench and pouring fire down at those who had not yet fled. The fourth tank pressed forward, ready to drop its enormous stepping stone into K2.

Private Eric Potten, the machine gunner in tank F23, looked through the narrow aiming port, tilting his weapon downwards as his vehicle turned to run alongside one of the trenches: 'a lot of the Germans got down into their dugouts. We picked them off as we went along.'

Lieutenant Mestwarb, another of the 84th Regiment subalterns, ordered his men to fire at the tanks and throw grenades, 'unfortunately without making the slightest impression on this monster'. British aircraft flew low over the German lines, feeding the defenders' rising panic. 'The tanks made rapid progress, crossing the trench to our left and right and firing constantly.' It had dawned on Mestwarb that 'this was no longer a battle; this was a massacre.'

His comrade, Saucke, after hours of fighting, was coming to the conclusion that the K2 trench line was also untenable. They could see British tanks to their rear, heading up the gentle ridge towards Flesquières. They would try to use a communication trench that stretched towards it to get away. 'Few words were spoken. What could we have said? It was not yet every man for himself.' A little later, around midday, that last bond of military discipline dissolved. Saucke, and the small group of survivors with him, ran into a group of British soldiers clearing the trench. The Germans were taken prisoner.

Having crossed the German trenches, Lieutenant Dillon of B Battalion, the Tank Corps, took stock. The Mark IV crews were in exultant mood: 'terrific, [our] tail was right up, we'd had a tremendous success. We had the sort of "we told you so!" attitude. The front had opened up like a pat of butter.'

By midday on the 20th of November, gains of 10,000 yards had been made in places. The type of advances that cost thousands of lives, and took months to achieve, had been made in a few hours. Brigadier Elles headed back to the British lines to send a victorious signal to London. The day's bag would include 8,000 German prisoners and 100 guns.

This astounding success prompted peals of church bells back home and excesses of verbiage from Fleet Street. Following the awful losses of Passchendaele, which many blamed on the British commander-in-chief, the *Daily Sketch* welcomed 'Splendid News from Haig'. The *Daily Telegraph* announced 'Thousands in Pursuit – Tanks Shock to Huns'. Colonel John Fuller, more often known by the initials J. F. C. Fuller, the Tank Corps chief of staff who later became one of the world's leading advocates of armoured warfare, declared the attack 'a stupendous success', defining it as 'one of the most astonishing battles in all history'.

The gains achieved by mid-morning on the 20th of November were, however, to prove a high-water point for the Cambrai offensive. From around 10 a.m. tanks of D and E battalions had pushed forward towards the Flesquières ridge, outpacing their infantry. Captain Jake Wilson later recalled that once they began their climb onto the feature, 'we got two direct hits in as many minutes; the first smashed the left track causing us to swing to the right to receive a broadside from the second.' His crew bailed out, leaving just the driver, killed by the second strike, behind.

A German artillery section occupying a walled garden knocked out the Mark IVs one after another. They likely claimed more than 30 British tanks. At the very moment then of its first triumph, it was all too apparent that this new weapon was vulnerable to some of those at the defender's disposal. The centre of the British advance had been stymied, so forces pushed on around it.

When the infantry and tanks reached the limit of their planned assault, preparing to hand over the exploitation to five divisions of cavalry husbanded for the purpose, the 'medieval horse soldiers', as one tank officer called them, had abandoned their advance after taking light casualties. Lieutenant Dillon, having fought through the enemy defence lines, remembered: 'everybody was absolutely furious that the road was open and no cavalry appeared.' Worse was to come.

In the days that followed, the Germans, fearing a spectacular breakthrough in the Cambrai sector, reinforced it, mounting a series of counter-attacks. Having exploited into the virgin lands beyond the trench lines, British infantry lacked its own defences from which to resist these onslaughts. The tanks could only help here and there because within a couple of days the great majority had been knocked out, suffered mechanical failure or got stuck.

As British losses piled up, and the gains of the 20th of November were whittled away, the inevitable recriminations began between senior British officers. At the troop level there was cynicism and disgust – Lieutenant Dillon describing the battle as 'typical of Haig . . . a fiasco'. I would rather avoid the whole 'Haig: Hero or Villain?' debate, but he and the other top brass stood accused of launching an attack that was too large to be considered a raid, too small to be an attempted breakthrough. Most importantly in the latter context, reserves of infantry, and indeed tanks, had not been prepared to reinforce the initial success, and to withstand the inevitable counter-attacks.

Surveying the knocked-out Mark IVs, and the charred bodies

of their occupants, one German general was appalled by it all. His verdict on tanks: 'I do not regard these things in their current form as battleworthy . . . however, perhaps they can be improved.' Even as it notched up its first triumph, we must record that there never was a moment in the tank story when it was invulnerable to enemy fire, and that the doubters were there from the outset too.

These immediate impressions, however, were recorded by German commanders humiliated on the 20th of November in a spirit of unbridled relief, exuberance almost, that they had been able to regain the lost ground. They were in fact part of a conservatism and disdain for this new form of warfare that would ensure that Germany failed to respond properly to the tank challenge, initially at least.

More thoughtful members of the German hierarchy came to different conclusions about what had happened. Field Marshal Paul von Hindenburg, while reasserting the primacy of traditional methods of fighting, might have been channelling the experience of those soldiers who met the first assault in November 1917 when he wrote: 'It was not so much that the machine gun and small calibre shellfire sprayed from these colossi was physically destructive; rather it was their relative invulnerability that was damaging to morale. Against their armoured walls, the infantryman felt himself powerless.'

So how did the British develop the Mark IV, bringing hundreds to bear at Cambrai? And who might claim the credit for it? That, after all, explains the origin story of an invention that came to symbolize warfare in the machine age.

Britain took to the tank. Its first battlefield appearance in September 1916 had yielded disappointing results, but no matter. War correspondents had waxed lyrical about 'toads emerging from the primeval slime' and crossing no man's land, and at London's Gaiety Theatre a cabaret singer, backed by chorus girls moving in formation as the Landship 'Tanko', sang: 'I'll teach the Kaiser and Co., we know how to win, when they see us dance the Tanko right into Berlin.' Those selling war bonds found that putting a tank in the main square would soon bring townspeople flocking in their hundreds or even thousands.

In wartime Britain, the tank married national pride at the country's innovation with desperation that there must be some better way to prevail than sending hundreds of thousands of soldiers to their deaths. For a nation traumatized by the losses of Loos or the slaughter of the Somme, the new invention literally seemed like the answer to millions of prayers.

One poster destined for a factory making them read: 'Every Tank turned out at these works saves the lives of our British soldiers'; another: 'The Tank is a travelling fortress that clears the way for our soldiers. It cuts through the wire under fire, it saves lives, it is *our* war discovery, it is a matter of pride to help to build Tanks.'

Such was the national excitement at this invention that in 1919 a Royal Commission doled out thousands of pounds to those who had brought it to life. As we will see, the evolution of the Mark IV seen at Cambrai is a distinctively British success story involving: great innovation; skilled engineering; Whitehall backbiting; and a stampede of people trying to take the credit.

Writing in the early 21st century it seems odd to see such a brutal weapon of war, spitting death from its guns and crushing any wretches who fell under its tracks, being extolled as a life-saving invention. Brigadier Elles would indeed describe it as 'a weapon for saving the lives of infantry'. People carried these thoughts forward, reminding everyone from the enthusiastic lieutenants leading their men, to the teenage factory girls who actually did much of the work building them, what the point of these new weapons was. The tank raised everyone's hopes of shortening the war – and no matter that its actual use on the battlefield had often achieved little, apart from educating soldiers that it could only be deployed successfully when combined skilfully with the other arms at their disposal.

On the outbreak of war in 1914 early dreams of rapid victory had, within a matter of weeks, transformed into a glum realization that to survive on a battlefield swept by shot and shell, men would have to burrow down into the soil. Dense entanglements of barbed wire soon became part of the picture – a way of preventing the enemy from racing across those tantalizing few hundred yards and surprising you at night, or, as the German defenders of the Hindenburg Line had trusted on the 20th of November 1917, stopping them from taking advantage of an artillery barrage to close the gap.

And as this stasis became, literally, more and more deeply entrenched, ideas of how to break through were tried and found wanting. Could you tunnel under no man's land? Or fly over it? Maybe just gas all the poor bastards in the enemy trench line? All of these tactics were debated, and many tried.

But for most of the war, the armies engaged in this battle of wills focused in the main on bringing ever greater quantities of artillery to bear (in order to cut wire entanglements as well as stun the enemy or kill him). When the Ukrainians disclosed in 2023 that they had fired 2 million shells in the first year of their war against Russia, First World War historians piped up to remind us that such quantities were used in a matter of weeks at the Somme or Ypres. A one-week preparatory bombardment had become a hallmark of General Haig's approach. And yet, still, the defensive mode of warfare almost always nullified the attack.

Colonel Fuller, the Tank Corps theorist, summed up the situation on the eve of Cambrai with his customary pith: 'the trinity of trench, machine gun, and wire made the defence so strong that each offensive operation in turn was brought to a standstill.' Just months into the war, British officers and inventors were, however, starting to break this problem down into its constituent parts in order to counter the trench, the wire and the machine gun. Ideas had been put forward for a machine with giant rollers to crush the wire. And a Lincolnshire firm had built a prototype trench-crossing machine even before 1914 was out.

In June 1915, less than one year after the conflict had started, Colonel Ernest Swinton, sent out to France as a military observer, and having witnessed several costly disasters, petitioned the commander-in-chief, urging the building of 'armoured machine gun destroyers'. These vehicles would harness the power of the internal combustion engine to take forward soldiers under armoured protection, closing that tantalizing gap with the enemy, then neutralizing the weapons that so often cut down the British infantry.

Later that month Brigadier Lord Cavan, commanding the 4th Guards Brigade on the Western Front, wrote to Swinton urging him to press on with his ideas: 'I welcome any suggestion in this extraordinary war that will help to take an enemy's trench without a cost of fifty per cent of the leading company, and seventy-five per cent of that company's officers.' If the advancing machine could breach the barbed wire too, so much the better. 'This is a *certain* saving of hundreds of lives and a fat legacy to morale,' wrote the brigadier.

Colonel Swinton, then, was one of the first to see the importance of combining engines, armour and weapons to make a breakthrough machine. So it is unsurprising that in 1919 he petitioned the Royal Commission, making his case as the inventor of the tank. But four years earlier the colonel had been hamstrung, not least by his lack of engineering know-how and the fact that the army brass was distinctly lukewarm about the whole idea.

Instead, and at the time of his 1915 memo entirely unknown to Swinton, it was the Royal Navy that provided the vision and seed money that created Britain's first 'landships', as the new weapons were initially called. It might seem like the proverbial Whitehall farce that teams of different people were working away on the same problem in ignorance of one another. But evidently such work was very secret, and in 1915 the type of vehicle that would result from this research was still quite unclear.

Armoured cars, wheeled vehicles mounting machine guns, had been born well before the war, and during its early months the

Royal Navy had embraced them as a way of defending its naval bases. But the kind of weapon that some people had in mind to break the Western Front stalemate was quite different to that. What was needed, some influential voices in the Senior Service had concluded, was something like a warship, only on land.

So it was that Winston Churchill, as First Lord of the Admiralty, struck by the fantastic possibilities of such a machine, and in great secrecy, formed the Landships Committee. The type of thing they initially had in mind was described by Albert Stern, a banker who had volunteered for the Royal Navy's armoured cars and later became a key member of that committee, as a vessel 'designed to transport a trench-taking storming party of fifty men with machine guns and ammunition'. The weight of such a behemoth, once clad with armour plate, might easily reach hundreds of tons.

If this sounds like a steampunk fantasy, a mission to create some destroyer bristling with gun turrets, disgorging dozens of storm troops, then it should come as no surprise that science fiction writer H. G. Wells was another of those who later tried to claim the mantle of inventor of the tank. In a pre-war story he had prophesied 'land ironclads' of exactly that kind.

As if to underline the Edwardian eccentricity of this, the committee was headed by a man who rejoiced in the name Eustace Tennyson d'Eyncourt, who was director of naval construction. He became yet another of those who would claim to the Royal Commission four years later to have had a pivotal role in it all, though he was in effective charge for a matter of months.

But the idea of a breakthrough machine, an actual landship, soon hit the buffers. The questions had multiplied: How would it be powered? Wouldn't such a thing weigh hundreds or even thousands of tons? And even if it could be built, how would you stop it sinking into the Flanders mud?

There was a testbed machine created in 1915. It built on an idea already adopted for heavy artillery pieces, that of mounting swivelling pads on its wheels to help distribute the weight on soft ground. This type of belt-linked system of rotating 'feet' was dubbed a pedrail. A prototype 40 feet long and 18 feet wide was constructed, harnessing a pair of engines powering pedrails.

Even bereft of armour or weapons that would have added many tons to its weight, this vehicle proved so unwieldy that the Landships Committee soon realized it was pointless to go on. So what next? While the navy had been taking the land ironclad idea a little too literally, army engineers studying the problem had realized that machines created before the war for agricultural use might provide some answers.

An American vehicle, the Holt tractor, was one of the best engineered. The War Office bought some of them and in early 1915 staged trials to investigate its cross-country and gap-crossing abilities. The Holt used caterpillar tracks of linked metal plates in order to increase the size of its footprint greatly compared to wheeled vehicles, and thus spread its weight, reducing ground pressure to the point that it could go many places where an armoured car or lorry would quickly sink up to its axles. Having been designed to tow ploughs and other agricultural impedimenta, this American machine had caught the eye of soldiers who wanted to haul artillery pieces or other equipment to places away from paved roads.

There were other examples too. Two engineering firms in Lincoln had built traction engines to slightly different designs. One made by a company called Fosters used another design originally intended for American farmers. This type of tractor replaced the big rear wheels normally seen on these vehicles with tracks, keeping a pair at the front for steering. While the army tried out its Holt machines, the navy examined some of these others.

Fosters had already been in receipt of a good deal of Royal Navy money. It had worked on a prototype of a giant, wheeled, battlefield armoured vehicle that would have weighed in at 300 tons. The firm had also built a huge tractor to tow 15-inch naval howitzers. The Landships Committee knew that Fosters had the engineering expertise to get things done quickly. That was important, because the study of these different options for a breakthrough vehicle took months rather than years. There was a sense of urgency. No wonder, given the dreadful ticker tape of death and maiming of friends or relatives mown down by machine guns in Flanders. There was also a sense that others were working on the same problem, a competition that became a matter of national pride as well as an ingredient of victory.

Just as the Admiralty and War Office had initially been carrying out trials secretly from one another, so other nations, including France and Austria, independently started looking at the American Holt tractor. Could its engine, wheels and tracks form the basis for an armoured war machine?

Fosters had in the summer of 1915 built what it called the No. 1 Lincoln Machine, which set an armoured box on top of an American-designed tractor arrangement called a Bullock. After giant wheels, pedrails and trench-crossers, the No. 1 Lincoln was the first British prototype that looked recognisably like a tank. An armoured box was built over the engine and tracks. The Bullock arrangement was turned back to front, with the wheels used to steer the vehicle projecting from the rear. A machine gun port was set facing forward in the hull, and there was a plan to mount a bigger gun in a rotating turret on top of the hull.

It cannot be said though that the No. 1 Lincoln impressed anyone during its trials in September 1915. At one event, staged in Burton Park not far from the works, a VIP delegation came up from London hoping to see something inspiring. Colonel Swinton, d'Eyncourt and Stern were among the committee members who watched the vehicle's tracks jump off its wheels every time it tried to negotiate a trench. It was also unable to mount a 5-foot-high vertical obstacle, a requirement set out by the committee.

Evidently the No. 1 Machine would soon flounder if sent against German trench lines. The vehicle was not long enough for that, and assuming that it tipped forward into a shell hole or obstacle, the arrangement of its tracks meant it would have trouble climbing out.

Despite all these difficulties, those running matters in Whitehall had seen enough to appreciate that Fosters represented their best hope of getting a breakthrough weapon through to the front quickly. The firm in any case was soon at work on its next prototype, which became known as Little Willie.

In July, the navy Landships Committee and its army equivalent had discovered one another's existence and agreed to pool their knowledge, forming a joint body. Fosters' records noted the first order for a prototype landship on the 15th of August 1915. Its engineers were already hard at work tackling the myriad obstacles that stood between that idea and reality.

The birthplace of the world's first tank was an agricultural machinery business on the outskirts of Lincoln. At the outbreak of war, Fosters employed 350 people; by its end this had swollen to 2,000. Its reputation as a maker of traction engines, pumps and other agricultural plant was well established, and the Landships Committee had already seen the product of thousands of pounds' worth of Royal Navy orders.

Central to its success were William Tritton, the general manager, and Walter Wilson, chief engineer. And these two names would join all the others claiming the inventors' prize money in 1919. But as we shall see, Tritton and Wilson had more of a right to it than many others because of what they achieved, and the staggeringly quick way they did it. Indeed, it's hard to think of a greater contrast with the Ministry of Defence's more recent procurement debacles than what was done at Fosters in the six months after that initial order was given in August 1915.

Wilson provided the ingenuity and Tritton the driving force. Their challenge came in three stages: working out the right design; overcoming the obstacles – human as well as technical – of building it; and getting their customer to accept it.

Their partnership resulted from the war. Wilson, a Cambridge-trained mechanical engineer, had initially served in Royal Navy armoured cars, before being seconded to Fosters in 1915. Tritton had been running the firm since 1905. One of the Fosters fitters remembered that Tritton 'demanded absolute perfection'.

As he prowled the works, 'you could imagine that his mind was working away all the time.' Indeed, so focused was Tritton that he had told his fiancée that they would not marry until he had delivered the first tanks.

Little did she know it, but her passage to the altar was hastened by the inspiration of Walter Wilson at about the same time as the initial landship order was received. The engineer knew that the military requirement to conquer a 5-foot step was proving impossible for the No. 1 Machine. As for the 9-foot trench requirement formulated by the War Office, it was out of the question. It was a matter of physics: the length of the track and the height of the front wheel it curved around meant it was a non-starter.

Wilson proposed a radical design departure. If you built a bigger hull, rhomboid in profile, and wrapped the track all the way around it, the caterpillars would be much longer. And by curving that track up and over the top of the hull, you would allow the tank to climb an obstacle or step of more than 6 feet, twice the height the No. 1 Machine was capable of.

A few days later this proposed innovation was signed off by Eustace d'Eyncourt at a meeting with Tritton in Whitehall. Whether or not Wilson realized it at the time, his rhomboid design had prompted one of the engineering compromises that would become characteristic of armoured vehicle development thereafter, where solving one problem created others. Not least, ideas of putting a gun turret on top of this new vehicle were abandoned by Fosters – it would have been too high off the ground, and unable to cover targets very close to the vehicle.

Wilson's inspiration prompted other questions too. Given the difficulties they had been having with the tracks themselves, making much longer ones for the new design seemed to be asking for trouble. Time after time, the caterpillars fitted to the No. 1 Machine had broken or jumped off the vehicle's wheels. It weighed 16 tons, much more than the Bullock tractor for which they were originally intended, and the pressures generated by mud building up against them proved too much. The Fosters team decided to develop their own tracks, a key step given how important those devices are to defining the difference between an armoured car and what the engineers would soon be calling a tank.

It was while they were working on the new design that leading members of the joint committee came to Lincoln's Burton Park for the uninspiring No. 1 Machine demonstration of the 19th of September. But something else had happened that day.

Once the No. 1 Machine had all too publicly failed its test, Tritton took Colonel Swinton (who had formulated the need for 'machine gun destroyers' three months earlier) into a large nearby shed to show him something. When the covers were pulled off, a full-scale wooden mock-up of what would become 'Mother' or the Mark I tank was revealed. The design was radically different to anything he had seen before, incorporating Wilson's wrap-around track. Swinton was stunned, writing later that it was a 'stupendous achievement . . . dinner on the train on the return journey to London was a joyous occasion.'

Just days later, on the 22nd of September, Tritton had more good news to report: a successful trial of their new caterpillar design. To understand it, we need to visualize the track as a series of metal plates, each joined to its neighbours by a steel pin, and when connected one to the other forming an entire chain of links rather like a giant bracelet.

A track made of smooth plates wouldn't work, because it would just keep slipping off the tank's wheels. It was similar to the problem of keeping a railway wagon from jumping the rails. So Fosters developed what they called a 'shoe', welding one onto each link in the caterpillar. This shoe was a bracket that embraced the wheels as it travelled over them, keeping the track in place. These brackets meant the new Fosters caterpillars were far less likely to jump off those wheels (as had happened repeatedly with the Bullock track fitted to the No. 1 Machine). The shoe also fitted naturally into the cogged wheel, or sprocket, that transferred power from the engine into the tracks. A modified prototype had been fitted with the new Tritton tracks, emerging with full marks from its early trials.

'New arrivals by Tritton out of pressed plate,' he telegraphed the committee in London, using suitably roundabout language for a matter of great secrecy, 'light in weight but very strong.' He signed off, 'proud parents'. For Bertie Stern, the banker on the committee, Tritton's language was just right. Solving the track problem was so important that Stern believed this moment 'was the birth of the tank'.

Dull or obtuse as all the ironmongery – that business of shoes, flanges and pads – might sound, Stern was undoubtedly right that the development of Fosters' new track was a key moment, arguably *the* eureka moment in the development of this revolutionary new weapon. After all, almost every other part, from the combustion engine to armour plating and armament, had already been combined in the armoured car. It was tracks that gave cross-country mobility, opening up the ability to traverse shell-cratered no man's land, crush barbed wire, cross trenches and neutralize machine guns.

An order soon followed for Fosters' rhomboid design. The first one moved out of the workshops under its own power on the 7th of January 1916: it had taken less than four months after Swinton had gasped at the wooden mock-up, and just nine weeks after the design had been ordered. The twin inventions of the new vehicle shape and the caterpillar design had produced a model that was capable of meeting the army's needs. How did they manage to manufacture it in a matter of months?

Tritton and his engineers were nothing if not practical. The Daimler engine and gearbox used in the huge Lincoln tractor they had made for the Admiralty to haul its howitzers would do. Its six cylinders and 13-litre capacity produced 105 horsepower, which they reckoned sufficient to move a vehicle of 28 tons. The Daimler may have had the advantages of being familiar to Fosters. Fuller later wrote that it was 'a known quantity around which the rest of the detail was designed'. But its performance was adequate rather than superlative. With this power-to-weight ratio the vehicle had a top speed little faster than a brisk walk – 3.7mph. It couldn't go very far either, since the onboard fuel got used up in around 30 miles.

Of course, it was a machine designed to break through trench lines where just a few thousand yards might separate the start line from the rear of the German position. They were hardly expected to drive hundreds of miles. As the crews who would man it would soon discover, steering the machine required a flawless dance by four different crewmen.

When it came to armament, another off-the-shelf solution was chosen. The Royal Navy came up with a stock of quick-firing 6-pounder (57mm) naval guns. Each vehicle would have a pair, mounted in armoured boxes, or sponsons, on the sides of the vehicle.

These tanks, later dubbed 'males', had an arc of fire from straight ahead around 100 degrees to the flanks. Firing an explosive shell, it was reckoned the 6-pounders could make short work of enemy machine gun teams or small bunkers. The Mark I would also be fitted with machine guns, one pointing straight ahead and another in each sponson. A different version of the tank, armed with five machine guns and called a 'female', was later ordered too.

Driving and fighting in this machine required a crew of eight. Armoured vehicle designers learned over time that the larger the volume, each man inside occupying around one and a half cubic metres, the heavier the vehicle for any given thickness of armour. In the case of the Mark I, that steel plate did not exceed 0.47 inches (12mm) in thickness, which was just about adequate to keep out enemy rifle or machine gun fire – unless they used armour-piercing bullets. And yet, because of the vehicle's large volume, it still ended up weighing 28 tons.

Making something that heavy inevitably prompted questions about how to stop it sinking into soft ground. Tritton's track design featured plates 1 foot 8 inches (508mm) wide. It was hoped that making them this big would help the vehicle to keep going and avoid getting 'bogged'.

As with the choice of engine or the mounting of the armament, this was a compromise. Making the tracks much wider would have added yet more weight to the tank and created its own mechanical challenges. Like any armoured vehicle, its crews would come to discover what kind of ground could or couldn't take its weight by a process of trial and error.

Late in 1915 the question emerged of what the landship should be called. Colonel Swinton was very hot on secrecy, believing that the factor of surprise would be key to its use, so that preventing the Germans learning what was in store for them was a matter of life and death. Tritton was far more laissez-faire, believing it was impossible to stop his staff from learning what they were building. These views collided at some of the trials in Lincoln, which committee members up from London were sometimes horrified to see had become a spectator sport for Fosters workers and some of their family members.

On Christmas Eve 1915 the committee found itself discussing a cover name for the vehicle, given that the circle of knowledge would inevitably expand as the effort to deploy it geared up. Anticipating that Mark Is, big metal boxes travelling under tarpaulins on the railways, would attract curiosity, they kicked around ideas like 'container', 'receptacle' and 'reservoir'. Why not call it a water tank? In the end, Swinton wrote, 'the monosyllable "tank" appealed to us as the most likely to catch on and be remembered.' If the 22nd of September and the successful trial of the new Tritton track marked the birth of a viable armoured fighting vehicle, so the 24th of December 1915 was the moment at which the new weapon got its name.

By early 1916 the prototype Mark I tank was being readied for its trials. If the political and military elite were to be convinced of the value of this hitherto secret work, things would have to go flawlessly, and it would have to be done with some flair and showmanship. It was decided to bring the Mark I prototype to Hatfield Park, where the Earl of Salisbury had kindly offered the use of his private golf course. Specially constructed trenches and obstacles were laid out for the tank to cross, bunkers upon bunkers, as it were. And given that it was much closer to London, Stern was able to ferry some of the dignitaries up in his own Rolls-Royce, giving another opportunity for the sales pitch.

On the 2nd of February key ministers and generals watched the demonstration, including the Chancellor of the Exchequer, Secretary of War, First Lord of the Admiralty, Chief of the Imperial General Staff and Master General of the Ordnance. Critically also, representatives from General Haig's General Headquarters had travelled over from France, including his deputy chief of staff and two corps commanders.

The signal was given and the 'gigantic steel cubist slug', in Swinton's words, started chugging forward, negotiating craters, crossing a 9-foot trench and knocking down trees. Tank advocates beamed with delight as the event passed off without breakdowns or lost tracks.

Field Marshal Lord Kitchener, the formidable War Minister, who was an engineer by trade and a scientific soldier by repute, declined to be impressed. He remarked that it was 'a pretty mechanical toy, but without serious military value', noted that such a large, slow-moving object would be quickly targeted by artillery, and left early.

Some of his staff would later claim that Kitchener's behaviour was part of a deception plan, that he didn't want to appear *too* keen. Hindsight is that proverbially wonderful thing, and in one respect he was right. Tanks would indeed prove vulnerable to well-handled artillery. Critically, though, Haig's men at the demonstration were impressed, one general asking Stern, 'How soon can we have them?'

King George V was also swept along by the possibilities of the new machine. When he attended a demonstration several days later, he took the chance to peer through one of the small sponson hatches to get a look inside. Writing to Churchill (who had moved on from the Admiralty), d'Eyncourt exulted: 'the King came and saw it and was greatly struck by its performance, as was everyone else.'

Before the month of February was out, the army had ordered 100 tanks, 24 to be made by Fosters, and the rest in the larger Metropolitan factory in Birmingham. Such was the urgency of the commission that Fosters had to hand over the drawings and assist Metropolitan, a heavy engineering plant used to making railway and tram carriages, to set up production.

For Fosters, even building 24 created a fresh set of problems to be overcome, not least in finding the workers to do it. Money helped: with the big production order Fosters' sales soared from £187,000 in 1915 to £400,000 in 1916. But finding the right people remained a challenge. The volunteer spirit of the early war months had given way in January 1916 to the Military Service Act. Tritton, who had previously had to defend valued craftsmen against accusations of cowardice for not joining up, now sought guarantees from the joint committee that they would not be conscripted.

As for the many additional workers needed to build tanks, young women were recruited. 'It was all girls, no men except the charge hands,' Dorothy Hare, recruited at the age of 15, would later recall. Her job involved making the shoes that were at the core of the caterpillar track. Their shifts ran from six in the morning to six at night, when another took over, and 'the machines were never switched off.' A six-day week paid her 15 shillings.

It was noisy, dangerous work. The shoes 'were rough cast steel, you could cut yourself on them', Dorothy remembered. One time she had forgotten there was a stacked pile of those components behind, which collapsed on top of her when she backed into it. Factory girls were required to wear bonnets because the belts driving drills and lathes ran overhead in the workshops and might catch their hair.

Another female teenage assembly-line worker recalled: 'the Fosters men who didn't go off to war had a little enamel badge so that people would know they weren't cowards.' Those who eventually returned were given their jobs back and the young women, 'munitionettes' as they were sometimes called, all lost their jobs.

By the early summer of 1916 the last of those challenges facing

Tritton and Wilson, gaining customer acceptance, appeared to have been met. The orders had come in all right, but whether the army really knew what it wanted tanks for, how it would use them, was another question entirely.

During the early summer of 1916, production of Mark I tanks got underway in earnest. There were bottlenecks, though, that meant suppliers struggled to meet the schedule for early deliveries. The supply of engines was one – even in November 1916, Daimler could only deliver 20 per month. Steel would prove a scarce commodity, given the vast scale of war production from shells to warships and tanks. And, as we've seen, Fosters and the other firms making tanks struggled to get skilled workers.

It was in early June that the soldiers training to fight in them received their first production vehicle. The schedule was that 25 tanks be shipped to France by the end of July.

Many of those concerned with their early development felt strongly that the element of surprise would form an important part of their power. In February, Colonel Swinton had circulated 'Notes on the Employment of Tanks', the first proper attempt to put forward what modern armies call doctrine. The point of the new weapon, he asserted, was 'assisting attacking infantry by crossing the defences, breaking through the obstacles and of disposing of the machine guns'. There was no point trying this with 'driblets' of armour; Swinton argued that a mass of 90 tanks should be considered the starting point.

So began the first of several doctrinal disputes about how best to use the new weapon. Haig, planning a great summer offensive, told Swinton in April that the planned deployment of tanks to France 'was too late – 50 were urgently required for 1st June'. Swinton was dead against this, but a colonel battling a field marshal needed allies.

The tank committee enlisted the support of Maurice Hankey, the War Cabinet secretary, who wrote to the Chief of the Imperial General Staff later in April: 'Sir Douglas Haig should be asked to do all in his power to avoid being committed to anything in the nature of a decisive infantry attack until the caterpillar machine gun destroyers are ready . . . a very large amount of money has been spent on them, and a great number ordered.' Hankey imagined a scenario in which the coming offensive started before the tanks were ready but as they began to arrive in France, an event that would be discovered by the Germans, 'by which time all prospect of their employment as a surprise would be lost'.

An appearance at Fosters that June added another level of complexity to the 'when and where' question. A French officer, Colonel Jean-Baptiste Estienne, inspected the work in Lincoln before travelling to see the first tank crews training at Elveden in Suffolk. The colonel, a brilliant innovator of nimble mind, reflected upon what he had seen. His thoughts on British tank design spurred ideas of his own, as we will see in the next chapter.

For the pioneers – both the Whitehall warriors and the soldiers preparing to use the new machines – Estienne's visit was a moment of revelation. Just as the War Office and Admiralty had discovered the existence of their rival projects the previous summer, now they learned that France had been working secretly on its own tank programme. The French, using Holt tractors as a basis, were developing an armoured vehicle armed with a 75mm cannon, the Schneider. Estienne reported back that the British were ahead. There were representations from Paris: surely Britain should wait until its ally was ready, so that the new weapons could be used simultaneously along the Western Front, compounding the shock to Germany?

On the 24th of June, one week after Colonel Estienne returned to France, the British began an enormous artillery barrage. By the time the whistles went on the 1st of July, and the men went over the top, more than one and a half million shells had been fired. Haig was not going to wait for tanks – not for British let alone French ones. There were hundreds of thousands of troops in play, and this was a weapon whose value was not even clearly established in his mind.

So began the Battle of the Somme, that dreadful confirmation of the bankruptcy of army tactics. On the 1st of July, the British army suffered more than 57,000 casualties, of whom 19,240 were killed, the worst loss in its history. Week by week the casualty rolls swelled, Haig's temptation to use tanks as soon as they were available growing with them.

Late in July the head of the army wrote to Haig: 'the use of tanks in small driblets will militate against their eventual value.' If the Germans had time, the letter went on, they would organize their anti-tank defences. The commander-in-chief in France faced other hurdles too. Neither the factories nor the army were able to meet the delivery schedules discussed earlier that year. It was not until the 22nd of August that the first Mark I tanks arrived in France.

For those crewing the tanks, every week of preparation time was precious. In keeping with the secrecy surrounding the project, recruits to what became the Tank Corps had been gathered under the cover name of Heavy Section, Machine Gun Corps. Early efforts to find mechanically minded men for the job led to adverts being placed in *The Motor Cycle* magazine. When trawling for volunteers who were already serving, only the scantiest details had been offered.

The units assembled in the French town of Bermicourt therefore contained a mixture of adventurers, those who liked tinkering with engines, and some who had understood what the next great step in warfare might involve. Fuller dubbed them a 'band of brigands'. Those detailed to crew the Mark Is soon understood that they were in for a horrible experience. They were crammed in around a throbbing engine, parts of which glowed red hot once it had been running for a while. 'Eight men in that tank altogether, not much room for dancing,' Lieutenant Norman Dillon joked later.

Driving it required the close cooperation of four men: the commander sitting at the front controlled the brakes; the driver next to him the throttle as well as the rear steering wheels on the

Mark I; and behind them two 'gears men', one for each track, who would use levers to select one of three options – slow, fast or reverse. Since the Daimler motor made so much noise, coordinating these actions was often done by hand signals and bashing a spanner against the armour: 'two raps for bottom gear, one rap means top gear.'

Private Jason Addy realized, early in his training, that 'conditions within the tank were appalling, not only for the noise, but also for the smell of the engine, exhaust fumes, cordite, to say nothing of the incredible noise and heat.' Bereft of any suspension or dampening, the men could feel every object they ran over, jarring and thumping. When crossing obstacles, the tank pitched forwards or back. 'If the hill is steep enough they may even find themselves lying flat on their backs or standing on their heads,' Captain Richard Haigh later wrote, though he insisted all that being thrown about could be 'rare fun'.

Just driving about in a Mark I was an experience of sensory overload: noise, heat, fumes, vibration, shock, shouting and rapping. Carbon monoxide poisoning and fumes often triggered vomiting, and any man unfortunate enough to touch the exhaust manifolds as the vehicle pitched about could suffer scalding burns. And that was before they had even got close to the enemy.

Going forward to battle, the 'bus', as their crews soon started calling them, would be loaded up with stores. Ammunition for the guns would be kept close by in racks. Lockers were stuffed with 30 tins of food, 16 loaves of bread, cheese, tea, sugar and milk. Since wireless communication between vehicles had yet to be introduced, flags, a system of coloured semaphore discs and carrier pigeons also had to be taken on board.

Sharing adversity inside a tank quickly built a particular bond between the crew and differentiated them from the Poor Bloody Infantry. While many of the foot soldiers reckoned the vehicles were death traps, and would rather stay outside them, the Heavy Section men soon convinced themselves they had the better of it. It was not, Captain Haigh believed, 'that death was less likely in a tank, but there seems to be a more sporting chance with a shell than with a bullet'. For another infantry officer who volunteered for the new corps, there was a feeling that its mission 'was coloured with the romance that had long departed from the war'.

Haigh was both right and wrong. The casualty rolls would demonstrate that you *were* less likely to die in a tank than on foot. And that was because of the second half of his formulation, the 'sporting chance'. The weapons raking the battlefield could all kill or incapacitate the infantryman. But only some could harm those under armour.

The arrival of the tanks near the front prompted a tide of visitors. The Heavy Section officers, knowing support for their project was vital, would put a handful of tanks through their paces, staging mock attacks, crossing obstacles, generally showing off. When the Queen of Belgium later visited, carpet was laid in a tank for her benefit. This mechanical 'variety show' was performed at a cost in time, training and mechanical breakdowns. What's more, it did not always have the desired effect.

Lieutenant General Henry Rawlinson took time out from commanding the ongoing Somme battle to see the armour. He reckoned tanks could be useful at night, and for knocking out strongpoints, but felt 'some people are rather too optimistic as to what these weapons will achieve.' The Prince of Wales, serving on the staff of XIV Corps, wrote to his father: 'they are nice toys, and worth trying, but not to be in any way relied upon for success.'

Both Rawlinson and his boss at GHQ, Haig, had a nagging feeling that the whole project hadn't been fully thought through. 'We are puzzling our heads as to how to best make use of them,' wrote the first, while his commander-in-chief mused: 'we require to clear our heads as to the tactical use of these machines.' These uncertainties led them to sack the first Tank Corps field commander, and inspired advocates of tank warfare to redouble their lobbying.

Colonel Swinton, having nursed the project this far, and an arch committee man, sought to iron out the differences. He emphasized the new arm's role as subordinate to the infantry, saving soldiers' lives, easing their passage forward. More radical spirits like John Fuller, who became the intellectual powerhouse of Bermicourt after his arrival that August, begged to differ with Haig's whole system of warfare.

Over the coming months (but before Cambrai) Fuller would take issue with key aspects of Haig's way of war. The week-long artillery bombardments turned the ground to pulp (making it very hard for tanks to get across), were so well understood that the Germans could set their watches by them, and worst of all, as at the Somme, hardly achieved any result.

Furthermore, Fuller believed GHQ was far too conservative in its use of tanks – better to focus them on the second and third lines of the enemy defence than the first. 'Though the Germans gave us a lot of trouble,' Fuller complained after seeing his ideas repeatedly thwarted, 'Sir Douglas Haig and his phenomenally unimaginative General Staff gave us infinitely more.' Fuller and Swinton at least agreed that there was no point trying tanks unless you threw large numbers of them into battle.

On this question, though, in September 1916 Haig chose to ignore the advice of the tank experts, the Chief of the Imperial General Staff, Churchill, other ministers, and his French allies. On the 15th of that month, he committed tanks to battle for the first time in history, in the Flers–Courcelette sector of the Western Front between Albert and Bapaume.

Just 32 Mark Is moved forward that morning, supporting an attack by 4th Army, an organization of hundreds of thousands of troops. These vehicles were parcelled out, ten to the Guards Division, four to the New Zealanders, and so on. First-hand reports speak of detachments making their way forward, some tanks bogging down, others failing, arriving in their ones and twos in no man's land.

Like many a Western Front battle, this first use of tanks ran the arc from hopeful initial reports of gains made here and there, to a gradual reckoning of loss and disappointment. German soldiers had managed to knock several machines out, in some places coming close enough to fire through observation ports or lob grenades onto the floundered monsters. The mud, churned by so many shells, proved a particular enemy that day.

Haig, who had learned to find justification in the most dismal of results, felt the experiment had been worth it. According to Bertie Stern, the commander-in-chief told him: 'wherever the tanks advanced we took our objectives, and wherever they did not advance we failed.'

Press and public reaction was more ecstatic. Freed at last from censorship, military correspondents filed reports from the front about the astounding new weapons unleashed on the Hun. To what extent Haig really saw the tank's potential and how far he allowed himself to be carried along by public and political enthusiasm, as well as the lobbying of enthusiasts within the army, remain moot.

Certainly, in the month after Flers–Courcelette GHQ in France wobbled, initially backing an order for 1,250 tanks, then cancelling 1,000 of those, arguing that the weapon had not yet proved itself. After uproar from the tank lobby in London, the bigger buy was reinstated.

During the months between that first use of armour and the success at Cambrai there were further attempts at attacks, the difficulties finally combining to convince even GHQ that Fuller's arguments about curtailing the pre-attack artillery bombardment in order to spare the ground and wrong-foot the Germans, as well as the need for mass employment of armour, were sound.

That period was also used to make changes to the tank itself. The evolution from Mark I to Mark IV made little difference to its outward appearance. Wilson's rhomboid design was kept. The most obvious change was in discarding the two-wheeled steering arrangement at the back. Although it still took four men to steer a Mark IV (and indeed to start its engine), modifications enabled it to be done without the wheels.

Crews in the Mark Is had suffered wounds from flakes of metal, or 'spall', that flew off the inside of their armour when the outside face was struck by bullets. The Mark IV was given slightly thicker armour (adding a ton to its weight) and chain-mail face masks introduced to protect the crew, particularly their eyes, from spalling. Some hard lessons had also been learned about vehicle survivability. The fuel tank, internal on the Mark I, was armoured and moved to the outside rear of the vehicle. The first Mark IVs reached units in France in April 1917.

With production underway of more than 1,200 Mark IVs, more factories across Britain joined the effort. The whole thing had grown way beyond the capacity of Fosters, which could only work on 24 vehicles at a time. By the end of the war, the firm had managed to produce just 101 Mark IVs.

It was a feature of Fosters' work that they only made the 'male' variant with 6-pounder guns, and, as production geared up, the army wanted more of those armed solely with machine guns. It was only towards the end of the First World War, when German-made tanks finally made their appearance and the first tank-to-tank engagements happened, that commanders realized the value of keeping male tanks in the mix. The ones armed just with machine guns could not destroy enemy armour.

The firm's designer Walter Wilson did not remain idle. A lighter, faster tank, the Whippet, was also built there. The idea was that they might be suitable for exploiting the gaps opened up by the heavies. Tank Corps officers had also started to think through the need for other variants – to carry supplies or infantry forward. Wilson started work on a 'gun carrier' to move an artillery piece. These vehicles, 50 of which were made by a neighbouring firm, mounted a 5-inch (127mm) howitzer. These were intended for use against more distant targets, including those beyond visual range, becoming the world's first self-propelled guns. After Cambrai, the evolution of the basic Fosters tank design continued with the Mark V, which had thicker armour, a more powerful engine and, at last, a different transmission, making it easier to drive.

The Mark IV was therefore an evolutionary milestone in the

dynamic story of early British tank development. Its significance was that it was the most numerous variant, and that it had a central role at Cambrai, rightly feted by British and German experts alike as the first successful tank battle in history.

In October 1919, the Royal Commission on Awards to Inventors gave its view on various claims to have fathered the new weapon. Winston Churchill, who certainly had a role, realizing the potential for the new machines and forming the Landships Committee while at the Admiralty, liked to be thought of as instrumental in the story. But he declined to submit a claim. 'Mr Winston Churchill has very properly taken the view', the Commission reported, 'that all his thought and time belonged to the State and that he was not entitled to make any claim for an award, even had he wished to do so.'

Others, including serving officers on the Whitehall committees, were less retiring. Even so, the Commission rejected claims by four of them outright. It did, however, make an award of £1,000 to Major General Ernest Swinton. As a professor of military history post-war, he also sought to cement his version of events by publishing a history on the 'Genesis of the Tank'.

And what of the banker Bertie Stern? In the view of one French officer who during the war liaised closely with the British on armaments matters, the main credit belonged to Stern because of his drive and vision. The Commission awarded Stern – who, like Swinton, had been a serving officer – £1,000.

In the panel's judgement, it was William Tritton and Walter Wilson who deserved the greatest plaudits for the invention of the tank. They jointly received £15,000. Tritton gave his half, £7,500, to Fosters' workforce. Not many managers of today would do the same. In his view he had already earned enough pay and had also in 1917 received a knighthood. Wilson invested his half in projects, including the development of a new gearbox.

The staff responded to Tritton's generosity by organizing a dinner in his and Wilson's honour in December 1919. The festivities and toasting took place at Lincoln's White Hart Inn, which had, four years earlier, served as a makeshift office for the design team. The staff lionized the two guests of honour as 'originators and designers of the fighting tank'.

It would be churlish to argue. As for the most successful tank to have appeared during the First World War, that was more of an open question. Because that French colonel who had visited Fosters in June 1916 would play the key role in developing one that would vie for those laurels.

Renault FT

Entered service: 1918
Number produced: 3,117 (in French factories)
Weight: 6.5 tonnes
Crew: 2
Main armament: 37mm cannon or 8mm Hotchkiss machine gun
Cost: Fr 56,000 (equivalent in late 1917 to £2,050; £119,000 in 2024)

Renault FT

Nothing was straightforward for the men of the Deuxième bataillon de chars légers, the 2nd Light Tank Battalion, that morning in May 1918. With the front giving way in many places, their unit had been ordered up the night before to stiffen French resistance. They had hefted their little Renault tanks onto a variety of trailers and flatbeds for the move, but there hadn't been enough, so the officers found themselves scraping together vehicles from the battalion's different companies in order to attack.

The tanks, weighing 6,500kg, each crewed by just two soldiers, were much smaller than the Mark IVs used at Cambrai by the British. And that was supposed to be one of their advantages – that they could be moved quickly by lorry or rail to points where they were needed.

And at this point they were sorely required. Germany's *Kaiserschlacht*, the 'Emperor's battle' or Spring Offensive of 1918, launched on the 21st of March, had turned the tables on the Western Front in dramatic fashion. In spite of all the debate about tanks and what they might achieve, the stalemate had been broken by brute force: 50 extra divisions of infantry freed up by Russia's capitulation the previous December, thousands more guns, 1 million shells fired in the first five hours of the German bombardment.

Into this maelstrom chugged these diminutive tanks. They were the brainchild of the architect of France's armoured forces, a vision he had pursued in the face of doubt and obstruction. Hostility in certain quarters of the general staff added yet more jeopardy to the moment facing that battalion.

In the opening phase of the Kaiserschlacht, Germany had sent the British 5th Army tumbling back in disarray; then they had turned their attentions to the French to their south, ploughing through their defences between the Aisne and Marne rivers, in places driving them over 35 miles.

As the 2nd Light Tank Battalion finalized its plans that morning, officers conferring near the calvary in the village of Dommiers, the scenery around them bore little relation to the sterile landscape contested over the previous years. The battle-lines had shifted suddenly. Farm buildings were intact, the trees verdant and the uncratered fields billowed with wheat.

They were supposed to be supporting the infantry in a move eastwards towards a suspected German advance – what military men call an advance to contact or encounter battle. Instead of the horrible certainties of two trench lines facing each other, this situation was full of unknowns.

In the end, the tanks moved off to the advance at 1 p.m. without the infantry. On the left, Captain Lemoine deployed in support of the 4th Tirailleurs but in fact well ahead of them, forming his Renaults in three sections, each of five tanks. To his left, one section extended in line abreast; to his right another, commanded by Lieutenant Aubert; just behind them and in reserve, a third. If you included the captain's tank, that made 16 Renaults moving into the wheat fields. On the right, the ground to their south, a similar group of 15 tanks was deploying in the sector of the 7th Tirailleurs but also well ahead of the North African infantry.

Lieutenant Aubert, peering ahead, anxious for any sign of the enemy, was worried about being spotted from the air. German aircraft had been sighted, as had observation balloons. But so far, the unspoilt countryside had masked them, 'crossing a brow crowned with wheat fields, the ears of which came up to the tank turrets, so masking the noisy echoing of moving tank tracks, that would not in other places have escaped being seen from above.'

The previous day, a German unit, the Württembergisches Gebirgs-Regiment (Württemberg Mountain Regiment), had advanced southwards along a road from the city of Soissons towards a village, Léchelle, about 4 miles to the east, and a little south, of the tanks' starting point. The two forces then were moving almost at right angles to each other, though the Württembergers, detecting the French on their right the previous day, had turned some elements to face them, and placed outposts of soldiers on that flank.

So it was that early that afternoon, emerging from the wheat fields, Captain Lemoine's leading sections, with ten Renaults, had successfully come through some of the German outposts without either side realizing what was happening. 'Suddenly, as if by magic, the enemy tank attack came,' the Mountain Regiment's war diary recorded. Machine guns chattered as the tanks moved through, mowing down surprised Germans on the way.

Now in open ground, Aubert recorded: 'we came under heavy machine gun fire, directed particularly at our vision ports.' Flakes of steel, spalling, flew off the inside of the armour on his left side, telling the young lieutenant where the fire was coming from. Having crossed the German line as he motored eastwards, the machine gun hitting him was due north. He fired, and then reloaded the little 37mm cannon himself; with just two crew, one man drove while the other had to do all the jobs in the turret. Aubert wound the turret around one full revolution until sure of his target – 'at less than 50m, with five shots, the machine gun was knocked out' – then directing the vehicle itself to run over the threat, 'it was comprehensively destroyed when the tracks crushed it.'

He had not long before seen a few German infantry surrendering behind him, the outpost line that they had driven through unawares giving up. In order to take advantage of the surprise they'd achieved, he needed the Moroccan Tirailleurs, but 'where was our infantry? They hadn't come through the last wheat field.' Aubert dismounted, going back into the field in an attempt to find them, bullets cracking and whistling about him. No joy, so after pausing their advance, the captain ordered the tanks on again, and Lieutenant Aubert mounted up. Moving on, the order was given by flag by the formation leader: Captain Lemoine's group of 16 Renaults advanced to the north of the town of Chaudun, the other group to its south.

Aubert, with Lemoine, was soon aware that they were being targeted by a German 77mm gun, well sited on higher ground to their front. He fired back with his own, smaller cannon, and 'as this duel went on a well-aimed shell landed in front of the tank, showering the crew through their viewing ports with its hot, dusty blast.' The Renault's shells may have made a lesser bang, but their smaller size allowed him to get shots off more quickly.

Most of the tanks to Aubert's left and right were equipped with machine guns, and they raked the enemy infantry as they retreated. The German war diary conceded that the regiment's 'unsupported wing was quickly bent back' by the French attack. However, the Mountain Regiment's account insisted that timely artillery barrages and close-range fire turned the tide. Either way, by around 3 p.m. the French attack had reached its limits. The southern group could not in any case advance into a gully ahead of them where the Germans were strongly posted.

The Tirailleurs and some French cavalry at last caught up, relieving the armour, so the captain gave the flag signal, 'turn about'. One by one the Renaults wheeled around, their tracks clanking, before heading back towards Dommiers, their starting point. Dismounting later, 'the crews [were] full of confidence and overjoyed at having accomplished their mission.'

By the standards of that war, the costs were slight: out of 62 men crewing the 31 light tanks, three had been killed, six wounded and two were missing in action. Command at times had broken down, with a couple of tanks, their commanders unsure what to do, just ploughing on regardless. One of these, reaching the southern gully, had been captured (that presumably explained the two men posted missing). A couple of other vehicles were lost to enemy fire or breakdown.

Aubert reflected that the section commanders, trying to retain control with flag signals from their hatches, while keeping situational awareness, were particularly vulnerable. Three out of the six section leaders taking part had been hit, one of them fatally.

As for what they had achieved, a group of little more than 60 troops equipped with light tanks had surprised and stymied an advance by a regiment at least 1,500 strong. The casualty returns for that day show that 20 or so Württembergers were killed, but there were captured and wounded too. Little wonder that Aubert and his comrades were exuberant. It was a most impressive debut for a tank that emerged as a partnership between one of France's most innovative military thinkers and the industrial know-how of a firm that remains a household name.

In the origin story of the tank, the record has long been rather Anglocentric. The British got them into battle first, it's true. What's more, at Cambrai, they did it at scale, changing military history. But the French high command was also actively discussing the need for armoured vehicles by late 1915, and it can be argued, as we'll see, that their armour played the bigger role in defining the outcome of the war at its decisive moment, the summer of 1918, when it seemed like the Germans could still easily win.

And again, while the story of French versus British tank research might involve any number of complex calculations or counterfactuals, one thing is clear: the British success was one in which the rush to claim credit produced comical amounts of self-promotion among rival contenders, whereas the birth of the French tank can be largely attributed to one man.

We glimpsed Colonel Jean-Baptiste Eugène Estienne (he preferred to be known as Eugène) during his short visit to Fosters' plant in Lincoln in June 1916. His bosses knew that he was the only officer who would be able to make sense of where the British had got to, for as Estienne later commented, '[the] English and French began to pursue their projects in the most complete ignorance of each other's work.' By the time of that visit the colonel was already in his mid-fifties and had been serving in the army for 32 years. An artillery officer by training, he had the gift of being both scientifically rigorous and possessed of a fertile imagination.

Before the 1914 war he had been involved in the establishment of French army aviation. He quickly decided that 'the aeroplane is the eye of the artillery', an observation that remains just as valid over eastern Ukraine in early 2025, even if most of those aircraft used today are unmanned. Estienne's study of the conflicts leading up to the First World War had convinced him that direct fire (shooting at things you can see) would be increasingly ineffective due to heavy entrenchments, and that measures were needed to improve the accuracy of indirect fire (at targets out of view).

During the early months of the Great War he served in the field artillery, and quickly understood that ways had to be found to overcome the Western Front stasis. He reportedly told his colleagues in the officers' mess: 'victory will go in this war to that of the two belligerents that is the first to place a 75mm cannon on a vehicle that can move on all terrain.'

By December 1915, Estienne was petitioning his commander-in-chief, arguing that 'it is possible to build motor vehicles that would allow the transport, across all obstacles, and under fire, at a speed in excess of 6kph [3.7mph], of armed infantry.' His initial ideas for a *cuirassé terrestre* or 'land cruiser' envisaged a vehicle of 12 tonnes, with a 37mm gun and two machine guns. Such vehicles, he believed, could tow armoured trailers carrying infantry.

Early film footage of Estienne shows a stocky man, thickened in middle age, but with a lively manner. His hair had turned white and was cropped almost to stubble on his big head, while atop his upper lip was a carefully clipped moustache. From the Lincoln visit onwards, he and the British cooperated closely.

In October 1916 the French tank force was formally established as the Artillerie Spéciale, with a salamander badge as its unofficial symbol. Estienne was promoted at that point to brigadier general. Central to the new force's mission was the support of those on foot, a later order defining it as 'accompanying artillery for the infantry, immediately acting upon the demands and necessities of combat'. Of course, Estienne was an artillery officer through and through, so the idea of giving mobility and protection to small field guns (37mm or 75mm) was central to his early idea. And his intelligence animated the whole project.

'Two things strike one immediately,' wrote Major Robert Spencer, who was posted to the Artillerie Spéciale as a British liaison officer, 'the charm of his perennial smile and the quick brilliance of his brown eyes. As a raconteur he is inimitable, whilst as a lecturer his marvellous power of expression, his command of vocabulary and his convincing use of simile make it possible for him to communicate to his less erudite audiences a certain measure of his vast knowledge.'

While the British boasted visionaries (Swinton or Churchill), gifted engineers (Tritton and Wilson) and tacticians (like J. F. C. Fuller), Estienne covered the gamut. Starting in early 1916 he was formulating ideas on how tanks could be used that were well ahead of those in Britain – or rather he was getting an agreed consensus among the French general staff on these ideas in advance of the likes of Fuller.

Estienne's approach could be summarized as: tanks should be used in large numbers across a broad front; they should not be employed over territory that had been heavily cratered by preparatory bombardment; ideally they would rely on concealment, using dawn attacks or smoke cover; and in a phased attack plan they could advance up to 6km deep, overrunning the enemy's field artillery.

Estienne had grasped the preconditions for a successful tank attack before Fuller, but that hardly endeared him to the acerbic Englishman. Indeed 'Boney', so called because of his diminutive stature and limitless ambition, who had an unkind word to say about almost everybody, wrote of Estienne: 'the general did not impress me although I found him an amusing little dud.'

As for France's tanks, Fuller described Renaults as 'nothing more than a cleverly made mounting for battalion machine guns', and the heavier model as 'a kind of kitchen range on tracks – unblushingly useless'. In this latter technical assessment perhaps Fuller's caustic phrases were a little more just. French tank development had been limited by many factors, and for as long as the war remained a contest to take deeply entrenched defence lines, they had struggled to come up with a design as successful as the British Mark I and its successors.

The Schneider, the first French tank to enter service, evolved from discussions that started late in 1915, 400 being ordered in February 1916, the same month in fact that Britain signed off on the production of the Mark I. British industry rallied to the cause rather more successfully.

While originally intended for delivery in November 1916, the Schneider order fell months behind. Estienne soon found himself struggling to keep the project moving – there were too many other calls on the scarce available resources, from orders for thousands of motor vehicles to pull artillery pieces to the phenomenal quantities of steel committed to the shells they would fire.

Estienne's battle throughout that year was with the generals and officials running the motor vehicle service. Captain Léon Dutil, one of the staff officers working for him, later wrote that British industry supported the tank project admirably, while advocates for the vehicle 'received from the army commanders nothing but opposition and distrust', whereas in France it was the opposite: army bosses believed in armour, those on the home front not so much.

The Schneider tank, when it eventually appeared, was never likely to impress. By mounting a metal box on a Holt tractor chassis, its designers had ensured that it would struggle to climb any significant vertical step. Although long enough to cross the trenches at the time of its design, the Germans started digging them significantly wider after the British tanks made their debut in 1916, making them too big an obstacle for the Schneider. In an attempt to deal with this basic problem they fixed a metal bar to the tank's nose, projecting out a couple of feet in front, in the hope that adding this 'beak' would effectively make the vehicle longer as it tipped into enemy trenches, helping it surmount the wider obstacle.

In April 1917 the Schneider made its lacklustre battlefield debut. At least the French had waited until they had numbers – sending 132 into action that day – and went to some effort to cloak their movements, trying to gain the element of surprise as well as shielding them from enemy fire. Even so, 57 machines were lost.

Next came the Saint-Chamond, a heavier vehicle (23 tonnes) mounting the hefty punch of a long 75mm cannon. Like the British tanks it had a crew of eight, meaning a large internal volume and rather thin armour, 11mm at its thickest. In a parallel of the British farce of teams working secretly without one another's knowledge, Estienne was unaware of the Saint-Chamond's existence for much of its development.

Had he been, the general might have pointed out the obvious drawbacks of its design: mounting all of this on a Holt chassis meant big overhangs fore and aft of the tracks, so when negotiating steep slopes (for example, shell craters or trenches), the nose or tail end would often dig itself into the ground, immobilizing the vehicle. That, and the heavy ground pressure that came from mounting so many tonnes on steel tracks that were too small, led it to be cursed by its crews as 'an elephant on gazelle's legs'.

Germany, which was far too slow to develop its armoured vehicles, eventually fielded a monster called the A7V, which shared many of the Saint-Chamond's faults. It was also based on a Holt chassis (it could only cross trenches 1.5 metres wide), weighed over 30 tonnes and had a crew of 18. Just 20 of them were made, and it proved even more prone to sinking in mud or getting stuck in shell holes than the big French design.

So, to the light tank, an idea that started to crystallize after the Lincoln visit in the summer of 1916. Estienne, seeing the size of the British machines, had come to the conclusion that a family of vehicles might be needed in the same way that artillery of various calibres had its different roles in battle.

What use did he think a little two-man vehicle would have? He believed they could be deployed in significant numbers, as mobile machine guns and light cannon to support infantry assaults. The light battalions, when they appeared, were organized in such a way as to plug neatly into the infantry formations they were supporting. The French light tank was therefore intended to give the closest possible help to those going forward on foot.

There are some suggestions that Estienne was inspired by the skirmishers of Napoleon's day, advancing ahead of the main infantry columns, picking off enemy troops and screening friendly ones. But if the Schneider, 6.3 metres long, had been too short to get across big trenches, how was the light tank, at 4.1 metres, going to do it?

Writing to the commander-in-chief in November 1916, Brigadier General Estienne argued that these vehicles would tackle such obstacles differently: 'it will cross shell holes, trenches damaged in places by preliminary barrages, not doing it like large tanks, bridging the gap, but going down into it.' The idea was to topple in nose first and climb up the other side. The eventual Renault design would be fitted with a large skid attached to the rear, to support its tail end as it powered out of the holes it had got itself into. Mounting a four-cylinder 40-horsepower engine, they trusted it would not just have the motive power to get through obstacles, but also move forward at a decent clip, up to 5mph.

To Estienne the mobility benefits came at the strategic level too. The light tank could be moved quickly on the back of a heavy lorry, on a trailer, or by rail. It could be raced to some threatened point, as it was destined to be in May 1918, or to exploit some success by friendly forces.

Perhaps the smartest win of the Renault design was that it took advantage of its small size and a two-man crew to enhance mobility while increasing protection. Encasing a crew a quarter the size of the Mark IV's meant a vehicle a quarter the weight, even though the Renault's armour was thicker, 16mm on its front. And, of course, both people and steel were scarce commodities by 1916.

By late that year the French had committed themselves to two tank programmes, neither of which had been proven in battle, and Estienne wanted a third. Many in the consultative committee, formed to bring together tank users and those who had to supply them, really couldn't see the point. Matters came to a head on the penultimate day of 1916 when a prototype of the new vehicle was presented to the committee at Renault's factory. It was a moment when Estienne's cherished project could easily have been snuffed out.

The ten members gathered there to see it included General Léon Mourret, head of the Direction du Service Automobile (the military motor service, responsible for making and operating the army's vehicle fleet), Brigadier General Estienne, Louis Renault, running the eponymous vehicle enterprise, and several staff officers. General Mourret in many ways personified the obstacles facing Estienne. He had to adjudicate between many different demands being made of France's automotive industry. But while battles between military users and procurement bureaucrats have become all too familiar in the defence business, Mourret was a particularly tricky opponent because his service was responsible for operating thousands of vehicles (including armoured cars) as well as making them.

Minutes for the meeting show that Mourret went in hard, trying to heavily modify or even kill off the light tank project. He couldn't see the point of designing a vehicle to mount one automatic weapon: 'it's not worth carrying a single machine gun, [it's] barely useful, and could go out of action without it being possible to replace it.' And what was the point of such a small machine more generally? 'He believes', noted the minute, 'simply that the solution put forward by M. Renault does not answer, that the 4t weight limit is unfortunate, and that it would be appropriate to give the constructor a bigger margin.'

The general's argument was that it was better to go either for a bigger tracked vehicle, or for more armoured cars instead. Monsieur Renault responded that the vehicle's small size was linked to the issue of simplicity and scarce resources. You could make it bigger, and modify the turret to house two machine guns, but he believed that 'if we modify the mock-up more profoundly we inevitably get to a weight of 8 to 9t.'

Brigadier General Estienne also defended the prototype. The minutes of the meeting note him arguing: 'it's essential to get the machine guns to the frontline. For [Estienne] the weight limitation of the equipment that will carry them is no less vital. What's more, the command that he represents has resolved itself in favour of the Renault vehicle.'

There were some other discussions about the thickness of the tank's armour and design of its turret. Making it out of riveted steel plates rather than casting it proved to be easier for the manufacturers.

In mid-January 1917, and after some further studies and exchanges of designs, the committee finished its deliberations. It voted by seven to three members to push ahead with the Renault design. Mourret was not entirely wrong, in that the new tank's weight did creep up, well over 4 tonnes, but Estienne had got what he wanted.

France's ability to make the new vehicle would depend critically on industry delivering – which in turn relied on another voice in that fateful committee meeting, and the second remarkable personality of this story: Louis Renault.

It was back in the summer of 1916, not long after his fateful visit to Lincoln, that Colonel Estienne bumped into Louis Renault in the corridor of a ministry in Paris. It was not their first encounter, the two having met at Claridge's hotel in that city late in 1915. Louis had been 38 at the time, was building a flourishing empire and radiated vitality. While Estienne's hair was white and his body portly, Renault had a thick, dark mane and luxuriant moustache, and wore finely tailored suits.

The army man had sought that Claridge's meeting because of the industrialist's great success in making cars and trucks. Renault declined to get involved with Estienne's project; the war was already giving him far more work than he could manage. But neither had forgotten the light tank project, and the encounter of July 1916 finally gave it decisive impetus.

Renault's father had set up a factory making buttons, of all things. But Louis and his brothers had been captivated by the speed and glamour of the motor car. So they started to make them in 1898 from a converted shed at the family home in Billancourt near Paris.

Renault Brothers, the car business, grew very swiftly, surviving the death of Louis' brother Marcel in the 1903 Paris–Madrid motor race, until by 1908 it was making 14 per cent of France's annual car production. Success on this scale brought enormous wealth, and Louis set up a bohemian ménage with a famous singer. Among their house guests were the composers Ravel and Fauré, sculptors, artists and actors.

Renault had been to the United States in 1911 to study emergent mass production methods and soon set about rethinking the way things were done at Billancourt. In racing as in the factory, the quest for speed obsessed him. 'Our days are numbered and our lives very short,' Louis later said. 'It's just human to try to discover, in the span of time we have, that which allows us to achieve the greatest horizons, to know the most, to see more countries, develop our thoughts, and experience the greatest number of sensations.'

Such was the importance of his car business to French life that Renault would become an incidental player in the Great War's two defining moments of flux. We have already glimpsed how the breaking of the linear stalemate in 1918 created the preconditions for his little tank to achieve big things. It was also true that soon after the war's outbreak in 1914, with German armies advancing steadily towards Paris, the French army halted them at the Battle of the Marne. And in this episode Paris's fleet of Renault taxis became part of a national legend. Army commanders commandeered 3,000 of the cabs to carry something like 5,000 reinforcements towards that decisive battle, so helping to save the capital. 'The Taxis of the Marne' became a symbol of national grit and solidarity – and no matter that the cabbies, who kept their meters running, billed the army for their services.

For Louis Renault the light tank project offered a chance to boost his military production even further. He was already making 5,000 artillery shells and 1,000 rifles a day. The range of products extended to aircraft, airship engines, components for artillery pieces, and of course trucks.

Billancourt, like Fosters in Lincoln, had increased production while seeing many key workers being called up. Out of a workforce of 4,500 workers in 1914, only 200 were exempt from conscription. Renault's answer to the labour shortage was to use young (French) women as well as workers brought in from African colonies. By the end of the war 21,000 people were employed at Renault's plants, nearly one-third being women.

This was manufacturing on a different scale to Fosters, and it would seem that from the outset it was the prospect of very large orders for the tank that appealed to Renault. Production would peak at 300 a month, with total army orders at one point totalling more than 7,000. In fact, delays and shortages meant that 3,177 of the Renault-designed light tanks had been delivered to the French army by the end of the war, and, like Fosters, the firm had to yield some of the assembly work to other companies. Even so, that more than 3,000 were made gives some idea of Renault's – and indeed France's – success in manufacturing.

The process of trying to gear up tank production took much of 1917. It could have been done much more quickly, but the arguments over priorities and shortage of materials, principally steel, held things back. It was also true that, once the vehicles started rolling out of the factories, a number of teething troubles emerged. There were particular difficulties with breaking fan belts. However, such was the appeal of the project to Louis Renault that, during the latter months of 1916, prior to the government tank committee giving production the green light, he invested the company's own money in developing the prototype. And he kept faith in 1917 as the army prioritized the manufacture of other vehicles.

The production engineer for the project was a Swiss working at Billancourt, Rodolphe Ernst-Metzmaier. He created a wooden mock-up of what would become a revolutionary design. Renault assigned a factory code to the tank, FT, which was no more than a project management convenience, despite some suggestions that it stood for *Faible Tonnage*, or, loosely, lightweight. But it led many to refer to the vehicle as the Renault FT.

Like Fosters' designers in Lincoln, Metzmaier and Renault used off-the-shelf components to speed the tank project forward. The 40hp engine had been developed for lorries, and the armament was standard weapons used by the French infantry. But Renault's experience with motor vehicles was so extensive that it was able to combine the engine with a simple, robust transmission to make the vehicle very easy to drive.

The driver, sitting at the front of the hull, had three pedals (as with cars, an accelerator, brake and clutch) and a bar on each side of him. These 'tillers' were adopted by many subsequent armoured vehicles, and allowed the driver to steer the tank very easily, pulling the left bar to stop the left track, slewing the vehicle in that direction, or the right one to go the other way. Compare that to the steering of early British tanks, requiring four men engaged in an elaborate performance.

Another aspect of the design that was decidedly superior to the British or early French designs was the enclosing of the engine, to the rear, in a separate compartment. This not only spared the crew from heat and fumes, but also allowed for ease of maintenance from outside.

The distance between that motor and the drive sprockets that would engage with the tracks in order to drive them round, was as short as possible, saving power and weight. At the front the caterpillars wrapped around a big wheel (known as an idler), before extending back towards the engine and drive sprocket. This large idler wheel, made of wood, helped the tank achieve a decent step-climbing performance for a vehicle of its size, though nothing like the British heavies.

When it came to the turret, the most obvious point was that it had one. While Fosters had just mocked up a turret on its No. 1 Lincoln Machine, the Renault FT was the world's first tank to be produced with one, giving the commander the ability to pan the weapon through 360 degrees, engaging targets wherever they might appear, or indeed firing while moving away from them.

One of Renault's early mock-ups featured a two-man turret. In the end, and after much debate, the decision was made to go for a single crewman. Adding space for someone else in the turret would have increased volume and weight, and of course required the recruitment of 50 per cent more crewmen. The disadvantages of having just one man in the turret would become apparent over time. It was too much to ask one person in the heat of battle to: navigate the tank; give orders to the driver; observe the movements of the rest of the section; spot targets; aim the gun, fire, and reload it. In this sense the Renault design would have lasting influence, but not in a good way.

Notwithstanding the rejection of the two-man turret, the vehicle's weight had crept up from 4 tonnes during the early design phase to 6.5 tonnes for the production vehicle. That was a function of many factors, including the decision to go for armour that was thicker than that on the French 'mediums' or British Mark IV. And switching from the 19hp engine originally considered to a 40hp one also added weight, even if it more than made up for it by producing more power.

Just as the question of crew size affected volume and so dictated weight, so questions of the motive force of a vehicle relative to its mass became vitally important for tank designers. The 'power-to-weight ratio', usually defined as the number of horsepower to the ton, became a critical measure of mobility. One doesn't need to factor in the difference between metric and imperial tons to see that the Renault with its ratio of 6.15 was quite superior in this regard to the Mark IV at 3.75.

This higher power allowed the Renault to move forward at 5mph – not exactly a heady speed but rather faster than the heavies. In trying to keep the size and weight down, designers accepted another compromise on the fuel tank. It only carried enough gasoline to drive for around 25 miles.

The very short range of the FT, and indeed the British Mark IV, underlines the limited horizons of those who made them. When one added together a move up to assault positions, advance to contact, fighting through enemy defence lines, with delays or pauses throughout with the engine idling, these early tanks were really just designed for a single day's action over a limited depth. As one of the Western Front's British commanders commented: 'they are not going to take the British Army straight to Berlin.'

At the time of the Renault production order in early 1917, the question 'What are tanks for?' prompted a variety of answers from senior military officers. The FT appealed to those who wanted a protected machine gun to support the infantry as they went forward – hence Fuller's characterization of the vehicle. At the time the project was launched, the trench warfare stasis had not yet been broken, so was there really scope for the Renault?

Estienne and his staff would have argued that it could get through dense defence belts and it is certainly true that they *planned* to equip the sixth tank in each light section with a bunker-busting short 75mm gun, and a droppable bridge attached to its front for trench crossing. But this variant of the vehicle was not produced in time to see action in the Great War.

Bereft of the gap-crossing variant, the Renault clearly was *not* a breakthrough vehicle capable of breaching heavy enemy defences like those encountered at Cambrai. But it was serendipitous for Estienne and Renault that just as production of the FT was gearing up and the army was forming its first light tank battalions, the battlefield situation changed significantly.

On the 18th of July 1918, France unleashed a large-scale offensive in the Marne area. Several weeks had passed since the delaying actions of the 2nd Light Tank Battalion, and now hundreds of thousands of French troops were to be engaged over the same gently undulating countryside near Soissons.

Although the front had moved little since that action at the end of May, the strategic picture had changed markedly. The window of opportunity seized by Germany to launch the Kaiserschlacht had closed. And while the initial effect of committing dozens of divisions freed up from the Eastern Front had been dramatic, the Kaiser's battle had lost momentum. On the 15th of July the Germans had launched a last attempt to push forward in the Champagne–Marne area. It quickly faltered, and the numbers were now on the side of their enemy.

A new ally, the United States, had joined ranks with the French and British. Two American divisions were committed to the Marne push, helping to bolster the battle-weary French, who fielded 22 divisions in this fight. They would also have the support of 238 tanks: 86 Schneiders, 85 Saint-Chamonds and 67 Renaults.

Although this was smaller than the number of British vehicles committed at Cambrai, the French ambition was greater: to drive back the Germans from their recent conquests in a multiphase battle. The French intended to push forward their initial assaults with the support of their bigger tanks, committing the Renaults to exploit success.

The attack delivered on the morning of the 18th was not preceded by a lengthy barrage. Rather, conforming to the ideas of Estienne and Fuller, there would be a short, sharp shelling of the enemy first line to keep their heads down as the Allied assault force moved up. Although more defensive trenches had been dug by this time, the ground was still relatively uncratered and the defences did not compare to those of the Hindenburg Line.

Indeed, those who led the advance recorded that the fields were still full of crops. This was hardly the sterile Western Front of Paul Nash's paintings.

Lieutenant Charles-Maurice Chenu, commanding a group of four Schneiders, described moving up in support of an American unit: 'I walked on foot, like the other battery commanders, trying to have, for as long as possible, a view of the group.' The vehicles chugged forward, soon crossing an initial German trench line without difficulty.

They weathered heavy machine gun fire that felled quite a few of the Americans advancing around them, before pushing on. Pressing on through the wheat fields, 'the reaper's blows', German field guns firing in the anti-tank role, knocked out two of Chenu's Schneiders in quick succession. On they went, with another battery of tanks, until they reached a ditch they couldn't cross, which was in fact their objective. There, sky-lined, 'one by one the shells found them, like balls in a game of skittles.' Those Schneiders were among the many knocked out by German artillery. During the first day of the Marne offensive, 61 tanks were destroyed.

Notwithstanding these losses, the first French attack wave gained ground quickly, the Germans being driven back a few miles in places. The Renaults were committed from the same wooded area that the 2nd Battalion had harboured in prior to May's action. Captain Dutil wrote: 'the appearance of French light tanks emerging from the forest of Villers-Cotterêts secured its eastern edges; the German troops were unable to enter it. An officer taken during the fighting argued these numerous light tanks produced excellent results; German artillery was unable to spot them; their sudden appearance caused great fear among the troops and inflicted serious losses.'

Estienne understood very well the tank's vulnerability to direct fire by German artillery, trying to shield them where possible or devote French cannons to support the armoured advance, firing smokescreens or neutralizing enemy guns. In bringing together what would more recently have been called 'the all-arms battle', the French were also ahead of the British in their use of a new technology: wireless radio.

Each light tank company commander rode in a TSF (*Télégraphie sans fil*, wireless telegraphy) version of the Renault FT. At Cambrai the British had resorted to carrier pigeons, but in the Marne battles the French had found a rather better way of coordinating their tank action with artillery and infantry. While the British Tank Corps also experimented with wireless in 1918, the French were ahead of them in employing purpose-built command tanks at low level.

Day after day the offensive was maintained, with the number of serviceable armoured vehicles dropping as it went. By the 23rd of July, just 85 tanks were engaged. But unlike the Cambrai battle, which began to unravel within hours of a dramatic initial success, the French kept going, and the Germans were unable to counter-attack successfully. The German army group war diary noted that French tanks were used 'in numbers never known before and much better developed technically'.

Blitzkrieg theorist General Heinz Guderian subsequently praised the French for the 'total surprise' achieved on the 18th of July and effectively credited the Marne offensive with turning the tide of the entire conflict. 'Ten German divisions were disbanded,' he noted; 'the High Command had to abandon the intended "Hagen Offensive" in Flanders. The Germans went on to the defensive along the entire Western Front, and the initiative passed to the enemy.'

Reflecting at a reunion of army comrades 15 years after the offensive, General Estienne waxed suitably lyrical: the 18th of July had been a 'glorious day'. The Schneiders and Saint-Chamonds 'opened the doors of victory that their younger brothers, light tanks, keep forever open'. The Marne battles and turning of the strategic tide solidified the position of Estienne and the light tank in the French narrative of success. He was lionized as 'the Father of Tanks', and the Renault as 'the Engine of Victory'.

The battles of July 1918 had also exposed the United States Army to tank action on a large scale. French mediums supported their initial assaults, and the Americans had decided to buy into the Renault programme, with production in the US eventually delivering finished tanks after the Armistice.

America's emergent armoured corps was provided with 144 French-built light tanks as a stopgap. Among the officers training on them in France that summer was a 32-year-old George Patton. In fact, he was the first person entered in the records of the newly formed US Tank Corps. Patton, a cavalry officer by training, gained swift promotion, running the first US Army tank school, commanding a battalion and then a brigade of FTs. He wrote the first US Army tank manual – which essentially aped French doctrine. In August 1918 at Saint-Mihiel, Patton notched up another debut, leading the first action in which American-crewed tanks fought.

Like some of the officers we've read about at Cambrai and Soissons, Patton went into action on foot. In one fight he clambered onto the back of an advancing Renault, but fell off, sheltering in a shell hole. By his exploits, being wounded and decorated during a few months of campaigning, Patton came to the notice of the press. One paper wrote up his deeds under the headline 'Col. Patton, Hero of Tanks, Hit by Bullet – Crawled into Shell Hole and Directed Monsters in Argonne Battle'.

Soon after the Marne battle, on the 8th of August, the British launched a major attack at Amiens. It was supported by 580 tanks, including Mark Vs (similar to the vehicles used at Cambrai but with improved transmissions and armour) and Whippet light tanks. The onslaught conformed to the surprise attack formula of a short, heavy artillery barrage followed by the advance of infantry supported by tanks.

Advances of several miles were made, and German morale shattered. General Erich Ludendorff famously described it as 'the black day of the German army'. The double Allied successes on the

Marne and at Amiens hastened the end of the war, and cemented the reputation of the tank, in both Britain and France, as a key ingredient of victory.

As these dramatic developments played out, some key players in the emergence of armoured forces began to organize themselves, anticipating that as the war came to an end they would face a new struggle. In the British army, 'Boney' Fuller was their standard bearer. In May 1918 he had presented his bosses at the War Office with a blueprint for the future, 'Plan 1919'.

In it, the colonel sketched out the revolutionary value of the tank, and its implications for the future shape of conflict. Armoured vehicles on tracks, far from sticking to roads, could cross 75 per cent of the battlefield, Fuller argued: 'he who grasps the full meaning of this change, namely, that the earth has now become as easily traversable as the sea, multiplies his chances of victory to an almost unlimited extent.'

He had started to think beyond the idea of the tank as a breakthrough weapon, something to end the Western Front trench nightmare, and to ask: what if we were just to keep driving? The British Mark V or Renault FT would not be suitable for the type of free-wheeling campaign Boney had in mind. They were too slow and only carried enough fuel to motor for a few dozen miles; thousands of new tanks would be needed, with markedly superior mobility.

However, November 1918's Armistice brought much of the armament industry to a clanking halt. From Billancourt to Lincoln, workshops fell silent, and, in time, the men who returned from the front took back their old jobs. There was no political will or money to build Fuller's new generation of armour.

Moreover, society had begun to absorb the collective trauma of the Great War, and those who remained in the military its lessons. Fuller and his acolytes were ahead of many, but almost all industrial countries had seen the role played by tanks, and understood how important it was for them to acquire such weapons. Among the vehicles developed during the late struggle, the Renault FT would become the outstanding export success.

France had produced thousands, many of which were no longer needed, so they were sold off to 16 countries, Poland being one of the first to send its new armour into action, against the Russians during 1919–20. Licensed production established in the US during the late conflict was curtailed, but after nearly 1,000 had been built.

Returning from Europe after the war, Colonel George Patton found reporters waiting on the quayside in New York. 'The tank is only used in extreme cases of stubborn resistance,' he told one scribbler. 'They are the natural answer to the machine gun, and as far as warfare is concerned, have come to stay just as much as the airplanes have.'

Armies across the world had reached a similar conclusion. In addition to the US, factories making local versions of the FT were established in Italy and Russia. In the latter case the Bolsheviks reverse-engineered their version from a vehicle captured from the Allied expeditionary force sent to fight them. From Brazil to Japan, Finland to Yugoslavia, the Renault became a sort of 'starter tank' for those who aspired to create modern armies. Indeed, working as a journalist in Afghanistan in the late 1980s, I was surprised to see one mounted on a stone plinth, guarding the entrance to an armoured brigade's barracks. The Afghan tanks were apparently taken from the Poles by the Russians, eight being presented to the Afghan king in 1923. It was claimed that one remained operational until the Soviet occupation of the 1980s, when it was photographed close to the Pakistani border. After the fall of the Taliban, a couple of surviving Renaults were spirited away to the US on board a C-17 transport plane; another returned to Poland.

Even setting aside the possibility that the little FT may have been operationally useful for more than 60 years, its record worldwide in the 1920s and 1930s was remarkable. Given the rather sedate top speed, short range and limited armament, one might ask why.

It was certainly the most practical vehicle to emerge from the late conflict. The British heavy designs saw limited use in the Russian Civil War (a British tank detachment having taken Tsaritsyn, later Stalingrad, at one point) and in some civil disturbances at home. But they were too cumbersome to be of much interest for any export customers. As for the Schneider, Saint-Chamond and German A7V, they gained a reputation for sinking in the mud or breaking down rather than winning great victories.

The Renault on the other hand provided a great starting point for emergent armoured corps worldwide. You didn't need to train many people, it was easy to drive, relatively simple to maintain, cheap to acquire and could get to many places that heavier designs couldn't. From George Patton to Charles de Gaulle, future tank commanders learned the basics on those vehicles.

The FT would also prove militarily valuable in the type of conflicts that many countries found themselves facing during the inter-war years. Spain and France both used the tanks against insurgents in North Africa in the 1920s. Whatever their limitations when facing a European foe, Renaults were immune to most things that rebellious tribesmen could throw at them. They would also play a significant role in the Spanish Civil War.

France held on to something like 2,700 Renaults through the 1930s, though only a minority of these were modernized and considered combat-ready when war clouds once again began to gather. The French had of course developed other tank designs prior to the outbreak of the Second World War, some of them very well thought of by contemporary experts.

But while France was swift and effective in its rearmament – adding around 3,000 more modern tanks to its inventory in the 1930s – something was missing. Estienne, promoted to major general, had accepted during the Great War that armour had to be completely subservient to infantry. In 1916 this doctrine had made eminent sense – it suited the realities of the Western Front, and helped to win over some generals who might otherwise have been obstructive.

Estienne remained true to this idea, telling the reunion of

X Army comrades in 1933, three years before his death: 'the only means of winning remains even today, as yesterday, the advance of infantry, and to work out the value of a new weapon one has to ask how far it will help that advance.'

As France was to discover in May 1940, the ideas and technology being developed across the Rhine had advanced far ahead of this. What's more, Germany had taken others' doctrine shamelessly, to produce a tank force that would conquer much of Europe.

Panzerkampfwagen IV

Entered service: 1939
Number produced: 8,553
Weight (Ausf. H variant): 25 tonnes
Crew: 5
Main armament: 75mm gun
Cost: 103,000 Reichsmarks (approx. £10,000 in 1940, which in turn is about £450,000 in 2024 pounds)

Panzer IV

The afternoon of the 13th of May 1940 was one of those times when all of the instruments in the orchestra of war came together. Looking across the River Meuse, Lieutenant General Heinz Guderian awaited the bombardment of French defences to his front. At 4 p.m., 'the battle began with a display of artillery fire which to us at least seemed magnificent.'

The French army defended the tree-lined banks of the Meuse from a chain of bunkers. Guderian watched them enveloped with smoke and dust as the barrage struck home, before hearing the drone of Stuka dive bombers overhead. His intention was to strike the enemy defence in great depth, bewildering its commanders and paralysing its forces.

At one point on the eastern bank a dozen or so *Panzerkampfwagen* IV tanks (it translates as 'armoured fighting vehicle', but Panzer IV hereafter) lined up after the artillery barrage had stopped. This vehicle, with a crew of five and a weight of around 20 tonnes, was a world away from the lumbering machines of the First World War. It was the most advanced vehicle in the inventory of the Wehrmacht or German military. These panzers hurled 75mm shells at the remaining strongpoints, providing pinpoint fire; some 88mm guns hit the ground targets too, joining in the cacophony.

Prior to the attack on France, after practice drills with their tanks, one of the German commanders had enthused: 'the Panzer IV is more effective on the battlefield than an artillery battery. Their firepower, if methodically and selectively used, can break any resistance.' Unlike the far-off gunners, these tank crews could see their targets, aiming directly for the slits of bunkers or snipers in windows.

On the 13th of May, the heroics were carried out by infantry, paddling across the river in assault boats. The bridges were down, and crucial hours would tick by before the sappers opened the first pontoon bridge at 11 o'clock that evening. But even before this happened, and panzers started to roll across, the German armour was exerting a powerful effect in the minds of the French defenders. The chief of staff of the 55th Division, holding this sector, found soldiers fleeing the assault: 'all the panicked men said that the enemy was in Bulson [on the French side] with tanks.' It was nonsense, a rumour.

The following day, the German armour started pushing outwards from the bridgehead. It was not a simple task, because this region of France, the Ardennes, bordering Belgium, was hardly the rolling plain imagined by many as 'good tank country'. Rather, the hills, steep slopes, river gullies and woods of this region channelled movement, giving the attacker very few options. It increased the tactical importance of barriers like the Meuse and pinch-points such as the town of Sedan, which, in theory, should have benefited the defender.

However, the German general staff had chosen to attack in this place because it allowed them to avoid the Maginot Line fortifications to the south, as well as the main concentrations of British and French forces deployed in the more open country to the north. Substantial German armies did advance against those Allied forces, but the three panzer corps, of which Guderian's was one, had been massed in the Ardennes region.

Thus, the plan concentrated these precious formations in order to execute a sweeping hook of an advance that would destabilize the entire Allied position. As the Germans made good their river crossing, sending tanks over the Meuse in the early hours of the 14th of May, the geography mattered less than the psychological effects of breakthrough. Encountering the invaders that day, one French colonel recorded: 'the enemy did not stop bringing tanks – about 60 came from Montimont between 13.00 and 14.30 hours . . . at 15.00 hours he was able to fire *simultaneously* at our blockhouses with *four tanks at a time*' (emphasis in original).

The French, aided by the Royal Air Force, tried their best to recover the situation by bombing the makeshift bridges. One air attack hit just as the army group commander, Colonel General Gerd von Rundstedt, and Lieutenant General Guderian stood in the middle of the span thrown across the previous night. French and British aircraft wheeled overhead, dropping their bombs, sending huge plumes of water into the sky, but the German flak batteries decimated whole squadrons. In the midst of this maelstrom, Rundstedt turned to Guderian, asking drily, 'Is it always like this here?'

Just as the air raids failed to sever the Wehrmacht's assault bridges, so French counter-attacks on the ground started to come unstuck. The first tanks encountered south of Sedan by the 1st Panzer Division were called FCM-36s. These were very much like the Renault tank of 1918, small machines crewed by two men, armed with a stubby 37mm cannon, albeit with thicker armour than the earlier vehicle. 'In a flash the company opened fire with every gun tube,' wrote a German tank officer, having sighted ten FCM-36s. 'The enemy was completely surprised. He did not fire a single round.' One vehicle was knocked out, others abandoned and three drove off.

As the battle developed, though, both sides discovered that effective armoured warfare would require a skilful interplay between their tanks and anti-tank artillery batteries. During the inter-war years, many nations developed specialist guns and ammunition to counter armour. These weapons were designed to punch through metal with solid shot. They ranged in size from the man-portable (just about) anti-tank rifle to larger guns weighing a couple of tonnes, mounted on wheeled carriages, with a metal shield giving some protection to the crew.

What anti-tank weapons had in common was that they were smaller and much more easily concealed than their targets. Even when you priced in a truck to pull the gun, they were also much cheaper. So, armies could afford to have far larger numbers of them than they could tanks, sprinkling the anti-tank guns along their frontlines covering all sorts of approaches that might be used by their enemy.

Faced with this proliferation of weapons, some compared the most effective tactics to the use of a sword and shield: advancing armour represented the sword, the means of attack, and anti-tank guns, skilfully hidden, often combined with other defence such as machine guns and mines, the shield. French troops facing the 1st Panzer Division had learned the hard way that standing firm with a group of light tanks simply invited their defeat. But in other engagements, French 25mm and 47mm guns, more easily hidden than an advancing tank, were able to 'shoot and scoot', knocking out the leading German machines before moving off to a new fire position.

This tactic was very well understood by the Germans too, and on the 15th of May, atop one of the low ridges to the south of Sedan, they were obliged to use it several times. Because within the wider story of the great German invasion sweeping into France, a subplot briefly played out in this place: a recently formed French armoured division attempted to throw the invaders back towards the River Meuse.

On the afternoon of the 15th of May, a French company of ten Char B1 tanks tried to advance towards the enemy, only to come under anti-tank gun fire, immediately losing two of their vehicles. The B1 was a formidable beast, considerably better armoured than the Panzer IV (its 60mm frontal armour was twice that of the German vehicle), packing both a 75mm gun in its hull and a 47mm, optimized for anti-tank performance, in its turret. These two very different tank designs came up against each other in the seesaw battle for the village of Stonne, just over 9 miles south of Sedan.

Several Panzer IVs had advanced into the village early on the 15th. The lead vehicle, bearing the tactical number 700, i.e. the company commander's tank, was immediately knocked out by a French anti-tank gun. The next one, tank 711, was also hit, a 25mm anti-tank shell having pierced the driver's position at the front, decapitating him. The four surviving members of 711's crew bailed out. Meanwhile, another Panzer IV burst into flames, destroyed by the French gunners.

Fighting raged and a couple of hours later some light tanks and heavy B1s of the French 3rd Armoured Division started to push their way in. The surviving crew members from Panzer 711 returned to their vehicle, engaging the attackers. 'It was a real monster,' the Panzer IV's gunner Corporal Karl Koch would later say of the B1. 'We fired 20 shots at it without success.'

The Panzer IV's main gun, a short 75mm weapon nicknamed the 'cigar stub' by its crews, was designed to drop relatively slow-flying high-explosive shells onto enemy positions. Guns designed to penetrate heavily armoured targets, on the other hand, fired a solid metal shot, a big bullet in effect, that travelled at a far higher speed. Even so, Corporal Koch managed to blow one Char B1's track off, immobilizing it. Another German gunner was able to find a weak spot, a radiator grille on the Char B1's side, causing a second French tank to explode spectacularly.

On the 16th the French renewed their attack, and Captain Pierre Billotte, commanding seven Char B1s, earned his place in the annals of French military history by briefly sowing chaos in the German lines before falling back. Billotte not only destroyed several enemy tanks during this raid but managed to keep his own operating for several more days, acting as a rearguard as French troops fell back.

Faced with these local difficulties, Guderian extracted his 10th Panzer Division from the area of Stonne, handing the fight to an infantry division. He bypassed the problem and pressed on with the advance of his XIX Panzer Corps, a great arcing movement across the north-east of France towards the Channel. 'I never received any further orders,' he would later write. 'All my decisions, until I reached the Atlantic seaboard at Abbeville were taken by me and me alone.'

What happened in France seemed to be the fulfilment of what the advocates of armoured warfare had been saying for decades:

that the mechanization of war would allow contests between nations to be decided far more quickly and at a fraction of the human cost of 1914–18. Guderian's XIX Corps, with 60,000 soldiers, over 500 tanks and thousands of trucks, embodied those theories, and never mind that much of the Wehrmacht still relied on horse-drawn transport. Instead, the point of difference, compared with the French army that attacked in July 1918, was that the panzer group carving its way across northern France relied more on machines and less on human sacrifice.

On they went, one German company commander recording: 'there was hardly any time for our tanks to pause: it was like being on an express train, one minute we are almost stopped, and then we are rushing forward again.' This is a key element of blitzkrieg, an advance so rapid that an enemy army with all its tiers of command, from battalion to brigade to division and so on, cannot react in time to stop it.

By the 24th of May, just 11 days after crossing the Meuse, Guderian's corps was laying siege to Calais. The Wehrmacht had humbled France, an astonishing reversal of the fortunes of 1918. One panzer officer, before going into battle, wrote: 'Today we will settle old scores; the French "gentlemen" shall experience our concentrated hatred.'

The fall of France marked the zenith of blitzkrieg. Its defining icons had all played their role, from Stukas dive-bombing command posts, to the 88mm Flak guns blasting away, and the dark grey panzers flooding towards the English Channel.

France had not neglected its defence – it had rearmed to meet the threat, manufacturing thousands of modern tanks and combat aircraft in the 1930s. In some ways its technology had been better. Certainly, the panzer crews had learned sobering lessons about the power of French guns and weakness of their own armour. As a design, the Panzer IV had been rather shown up. But the German system of war had proved markedly superior. Critically also, its army had been better motivated and trained. When it came to the Panzer IV knocking out a better protected and armed Char B1 tank, the human factor was vital. In the story of the tank, we will see that repeatedly: skill with the machine outweighing some of its inherent weaknesses.

German propaganda made much of the success of blitzkrieg and the swift humbling of France at such a low price in blood. And if this success were to usher in a new Reich, could its practitioners become a new sort of nobility? The 'Home Report' programme on German radio reported in June 1940: 'new tanks ready for attack, ready for a mighty push forward, these tanks carry with them the new romance of fighting. They are what the knights were in the Middle Ages.'

The Panzer IV was at this point a supporting player in the *Panzerwaffe*. But how, given the limitations it showed in France, does it justify its place in a selection of the most important tanks in military history? Firstly, because it really was far better than those vehicles around at the end of the First World War or in the 1920s. Secondly, it was to prove a very successful shape-shifter, evolving in ways its Panzer III brother could not, with the result that uniquely it remained in production from the first day of the Second World War until almost its last. It was central to the German panzer arm at a time when the Nazi leadership unleashed terrible conquests and global strife. As such it is an exemplar of warfare in the machine age at its height.

The development of the panzers that found themselves facing French armour in the Ardennes, and the doctrine behind their use, were features of the late 1930s. Germany had been banned by the Versailles Treaty, ending the Great War, from having tanks. General Guderian's resentment was evident when he later wrote: 'our enemies regarded the tank as a decisive weapon that we must not be allowed to have.'

Although Germany had been running some covert projects to develop armoured vehicles, full-scale breakout from the Versailles terms only occurred in 1935 after Hitler's rise to power. His forces had very little experience of armoured operations, or much idea of what their tank force should look like. General Ludwig Beck, the chief of staff of the German army, commented that same year: 'because our practical experience is insufficient . . . equipment and organization should be considered from countries that have gathered practical experience.' Guderian himself, writing a book called *Achtung – Panzer!* in 1937, was more explicit, saying that having rejected French doctrine, 'after mature consideration it was decided that . . . we should base ourselves principally on British notions.'

So it came to be that the panzer force that invaded France was partly modelled on British ideas of what a tank corps should look like. It was a bitter historical irony that these doctrines were picked up much more energetically from 1935 by Germany than by the British government, and fused in a highly successful way with other elements of blitzkrieg.

One of the main influences on Guderian, and his boss as head of armoured forces, General Oswald Lutz, was none other than that acerbic genius of the Royal Tank Corps, Colonel Fuller. His ideas informed both the German sense of what a panzer division was for, and the type of machines that would be needed for success.

In his 'Plan 1919' blueprint, Fuller had argued that the use of tracked vehicles on a large scale would allow for operations to develop much more quickly: 'Every principle of war becomes easy to apply if movement can be accelerated and accelerated at the expense of the opposing side.' In this war of movement, the paralysis of an enemy would require the attacker 'to destroy "command", not after the enemy's personnel had been disorganized, but before it had been attacked'. If all this reads prophetically in terms of what happened to the Poles in 1939 and the French in 1940, it proved far from easy to convince many conservatives in the higher commands of Europe, including in Germany.

While they had started forming panzer divisions in 1935, Guderian would later argue that the acceptance of the central place of these strike forces in his army's doctrine happened only after the replacement of the conservative General Beck in 1938. It was only in November of that year that General Walther von Brauchitsch, the army commander-in-chief, formulated new instructions for the Wehrmacht that spelled out the dominance of armoured formations. 'At the start of a war', he wrote, the panzer forces 'should be sent to operationally important positions, occupy strategically important sectors by surprise, or lead the mass of the army in decisive operations.' Tanks were to be the stars of the show. Guderian had won the argument.

That 1938 directive settled, as far as Germany was concerned, the debate about whether tanks should all be massed or distributed across the army. Other European armies had emerged from this debate with their own answers. The French had scattered much of their fleet in small detachments with the infantry. In 1940 that meant there were many tanks sitting idle in subsidiary parts of the front, while the Germans focused theirs at the key point.

In Britain, the theorists so admired by Hitler's generals had formed an 'Experimental Mechanized Force' as far back as 1927. It later became the 1st Tank Brigade. But while the British military had its radicals arguing their case that armour would transform the face of war, it also had its reactionaries who held back change, and a public that was still so traumatized by the Great War that it was very reluctant to divert huge sums to rearmament. When that finally happened, the British continued to buy tanks of different kinds (to support the infantry or move rapidly across the battlefield to exploit opportunities) and organize them accordingly.

In the Soviet Union, where some senior officers had also been influenced by Boney Fuller's ideas (and they'd been translated into Russian for widespread circulation), an unambiguous industrial commitment had resulted in a very large tank force. But its organization and direction had, like so much else in the Red Army, been disrupted by purges of the higher command. Only in Germany did the ideas, organization and machines come together on a grand scale by 1939.

History has been generous in giving Guderian the credit for many things, causing some to denounce him as an ardent self-publicist and a convinced Nazi. In many ways his rise, descended from generations of Prussian officers and enduring numerous scrapes in the trenches, was archetypal of those in the higher reaches of the Wehrmacht. But he had unusual qualities too, including a willingness to rock the boat in the cause of modernizing the army and to play politics to get his way. Publishing his book *Achtung – Panzer!* was just one example of his willingness to lobby publicly for change.

While Lutz fell by the wayside, in part because of insufficient zeal for the Nazi cause, Guderian, once he had met Hitler, was able to stay in his good books. Having worked in the staff on armoured matters as well as specifying the shape of the future tank force in the 1930s, then commanded a panzer corps in the invasion of France, he was in a rather unique position as both theorist and high-level battlefield commander, something neither Estienne nor Fuller managed.

Furthermore, there are eyewitnesses such as Albert Speer, later the armaments minister, who tend to bear out Guderian's description of the arguments around the tank force. Of course, Speer also faced accusations of falsifying history in matters such as his complicity with slave labour. But he is reliable on the more arcane issues, for example battles with the Führer over armoured vehicle design or production priorities.

As for the emergence of blitzkrieg doctrine and the development of the panzer arm, naturally Lutz and others in the 1930s had a share of the intellectual ownership. Similarly, Basil Liddell Hart and other British thinkers picked up the baton of developing armoured warfare – and his ideas may also have influenced the doctrine put forward by Brauchitsch – even if Fuller garnered more attention.

From a narrative point of view, though, it is hard to nudge Fuller off the stage since, having drifted into occultism and fascism in his later life, the former British colonel found himself standing next to Adolf Hitler at a huge military parade in Berlin in April 1939. Fuller had ignored Foreign Office requests not to go. He didn't intend to miss the spectacle, for the panzer phalanxes took three hours to pass. Having smarted with frustration at his own army's indifference to his ideas, Fuller watched them made a reality by the Nazis. When the parade had passed, Hitler seized Fuller by the hand, enquiring, 'I hope you were pleased with your children?' The British officer replied: 'Your Excellency, they have grown up so quickly that I no longer recognize them.'

The shape and purpose of those panzers were directly influenced by concepts Fuller set out in the late 1920s. He argued that an army needed distinct classes of armoured vehicles for 'finding, protecting, and hitting . . . in other words – reconnaissance tanks, artillery tanks, and combat tanks.' This concept was developed by other British pioneers (after Fuller left the army in 1927), with the medium-weight, faster vehicle being called a 'cruiser', and the heavier or 'protecting' sort classified as 'infantry' tanks.

Adapted to German doctrine, the ones intended to exploit breakthroughs would have high-velocity guns, albeit of relatively small size, firing solid shot to punch through enemy armour, while the infantry tanks would fire high-explosive shells in support of the foot soldiers they accompanied.

Lutz and Guderian thus defined two key machines suitably armed for their different roles: the Panzer III with a 37mm gun, and the Panzer IV with its short 75mm weapon. The latter was initially classified as a *Begleitwagen* or 'escort vehicle'. They joined the small Panzer II, with its rapid-firing 20mm cannon, which filled the reconnaissance vehicle role. These weapons had quite different characteristics.

As these theorists had seen in France, battle had a way of making a nonsense of such calculations. What if your light tanks ran into his mediums (as happened with the French FCM-36s wrecked by the 1st Panzer Division after it crossed the Meuse)? Or if the enemy started firing at your Panzer III medium tanks with well-concealed anti-tank guns, when actually the Panzer IV's gun was much better able to counter such defences?

It had occurred, naturally, to the theorists in the 1930s that the enemy couldn't be relied upon to be cooperative. There were various answers, such as organizing mixed units, so that vehicles like the Panzer IV might cover the Panzer IIIs from some suitable vantage point as they advanced, ready to neutralize enemy anti-tank guns. But what if that wasn't possible? Why not combine different weapons on the same vehicle? After all, you could have a secondary armament, like the machine guns incorporated in many late-1930s designs.

British designers had shown some confusion in this. Having given an 'infantry tank', the Matilda II, a high-velocity, anti-tank weapon rather than something firing high-explosive shells, they felt that a machine gun would be quite adequate for the 'helping the infantry' mission. In common with other nations, the debate between those who believed in specialized tanks versus advocates of a single vehicle able to fulfil a variety of missions went on for many years.

In trying to resolve it, some inter-war designers tried fantastical solutions. Why not mount one type of gun in the hull and another in the turret? The French army tried this with the Char B1 – the 75mm, firing its explosive ordnance, was in the hull, whereas the 47mm weapon in the turret was better at punching through armour plate. The Mark I version of the British Churchill tank similarly had an infantry support howitzer in the hull and an anti-tank weapon in the turret.

Or why stop at that? Why not have lots of turrets, offering a weapon for every tactical situation? The Soviet T-35 had five turrets. Two contained guns optimized for anti-armour work, a further pair had machine guns and the fifth a 75mm infantry support gun. It proved a nightmare to operate, weighed too much and still managed to have poor protection levels because of the volume needed by its ten crew.

Successful designs ended up with a single, heavy weapon for the main job, and one or two machine guns for flexibility. Thus, Lutz and Guderian had different tasks in mind when they drew up the operational requirements for their two new vehicles. Their template was copied from the British even though they were quite uncertain whether that was the right way to go. The contracts for initial work on the German version of an infantry tank had been awarded in 1935.

There were competing offers from Rheinmetall and Krupp, the latter soon emerging triumphant. In fact, the Magdeburg-based Krupp-Grusonwerke was duly awarded a production contract late that year, with the first completed machines made in June 1936. The initial batch or 'A variant' was only a few dozen vehicles but it represented a great step forward for the Wehrmacht, which had initially been training its battalions with cars and motorcycles, before adopting the 'training tank' Panzer I, a machine-gun-armed light vehicle designed to circumvent the Versailles restrictions.

What drove out of the factory in June 1936 was a vehicle that had been shaped through the interplay of ideas from those running the nascent panzer arm who defined the requirement, officials at the Heereswaffenamt (the army ordnance office or HWA) who drew up technical specifications as well as reviewing the results of trials, and the people at Krupp-Grusonwerke who would actually make it. The chief designer at that plant, an engineer named Erich Wolfert, was the father of the Panzer IV in the production engineering or technical sense. He had been involved in Germany's tank-making project from its start in 1929 when the *Leichttraktor* was developed in secret, because of the Versailles veto on tanks. Under the 'light tractor' cover project, Wolfert and the HWA experts he liaised with became trusted insiders in the birth of a modern panzer industry.

Years later, their ambitions had grown considerably in scale. The Panzer IV, almost treble the weight of that secret vehicle, used a 10.8-litre V12 petrol engine made by Maybach. Its tracks were driven by sprockets at the front of the vehicle, an arrangement that required power to be taken from the back of the hull to its front, losing a few horsepower along the way and eating away at space inside. The placing of gearboxes at the front of the tank also made them more vulnerable to damage. But Wolfert and the HWA accepted these compromises because the prototype of an earlier experimental machine (codenamed the *Grosstraktor*) that was driven from the rear often threw off its tracks. Arguably, the way the gearboxes were installed on the Panzer IV also allowed them to be more easily maintained, through hatches built into the hull front.

Whatever the compromises made on the transmission, Krupp's vehicle marked a dramatic improvement in mobility over the First World War designs. Whereas they might travel 20 to 25 miles before refuelling, the Panzer IV could go 80 miles cross-country, and exceed 125 miles on-road. By keeping the weight (initially) under 20 tonnes it had a very healthy power-to-weight ratio also, easily reaching 18mph. It really had been built to conquer Europe.

The Panzer IV's turret was crewed by three men: commander, gunner and loader. This was to set the pattern for other tanks later developed worldwide. It was adopted by the Panzer III too, both tanks also placing the commander in a central position and (from the B variant of the Panzer IV onwards) giving him a raised vantage point offering all-round vision through bulletproof viewing periscopes. The type of commander's station, termed a cupola, also became a standard feature of tank designs.

Its 75mm gun really had been designed to help the infantryman rather than knock out other tanks. High-explosive shells flew from its barrel at the speed of 420 metres a second, and it could reach targets as far away as 6,500 metres. The compromise here was obvious to the HWA, which even before production of the tank had begun noted in one of its reports, 'the low muzzle velocity results in a curved trajectory which will seriously affect accuracy.'

So, if you fired at an enemy tank a few thousand metres away, the 75mm shell, at that speed, would take several seconds, maybe even more than ten, to arrive. It was fine for hitting pillboxes or other things that stood still, but against an enterprising enemy tank crew who kept moving, they might finish their cigarettes, start up and rumble away in the interval between the flash of a shell leaving the Panzer IV and its arrival at their former location.

The trajectory, flying higher in the sky than the 37mm shot fired by the Panzer III, also reduced the chances of hitting something, particularly a moving target. Not only would the slower, higher-flying shell be more subject to gusts of wind, but also a high muzzle-velocity weapon, firing with a flatter trajectory, was less likely to over- or under-shoot the target.

So, Britain's 2-pounder (40mm) Quick Firing gun, fielded on an anti-tank carriage and as the main armament for the Matilda II tank the Germans encountered in France, had a muzzle velocity of nearly 800m per second. Its German equivalent, the 37mm fitted to Panzer IIIs, while designed for the anti-armour role, had disappointed in France, leading to a reliance on 88mm Flak guns to deal with heavily armoured threats.

But Lutz and Guderian to some extent failed to anticipate this when they specified these panzers in the mid-1930s. The idea that their design would be a dynamic compromise between firepower, mobility and protection was already well understood. Stretching that eternal triangle in the direction of mobility, as they had, produced significant consequences for the vehicle's level of protection.

The Panzer IVA was made with hull armour 14.5mm thick and just 20mm on the turret. It was immediately recognized as inadequate, so the variants used in France had 30mm in both places, but as the crews fighting near Sedan had discovered, that was much less than the modern French medium tanks. Also, the welded steel plates that made up the vehicle's hull presented vertical surfaces to incoming shells. The French made better use of sloping armour, a way of increasing the depth of steel that an incoming shot had to penetrate, without actually requiring thicker, heavier plate.

In a breathless piece of armoured troops advocacy for a German officers' magazine in 1937, Guderian had stated: 'If the army can in the first wave commit to the attack tanks that are invulnerable to the mass of the enemy's defensive weapons, then those tanks will inevitably overcome this their most dangerous enemy.' So in France did either of Germany's most modern tanks meet the test of being 'invulnerable' to most of the defenders' weapons?

This idea will recur again and again in this book. No tank is invulnerable, so how much protection is enough? If, by 'the mass of the enemy's defensive weapons', you meant their rifles, machine guns and hand grenades, then clearly a fighting vehicle can survive the basic threats. But what about more specialized anti-tank weapons? In May 1940 panzers had fallen victim even to the lighter-calibre ones.

Asked for feedback after the campaign, the division operating Panzer IVs at Stonne reported: 'armour protection is too weak and offers little or no protection against French 25mm and 47mm anti-tank guns.' It was a measure of the adaptability of the basic Krupp-Grusonwerke design that the series of up-armouring packages that helped add 7 tonnes to the weight of that tank did not seriously impact its mobility; rather, many engine upgrades helped the horsepower keep pace.

France had also shown that viewing armoured vehicle design entirely through the prism of the firepower–mobility–protection triad could not entirely explain the difference between winner and loser. One key factor – identified by some as a fourth variable in the design – was the flexibility offered by radio communications. This was important for everything from how one vehicle communicated with another, to the ability of an entire force, equipped with thousands of tanks, to adapt to sudden changes on the battlefield.

In the years since the First World War, there had been a dramatic change in the way that commanders of armoured units performed their role. Back in 1917 or 1918 quite a few tank officers, we've seen, went into battle on foot, some even on horseback. In the 1930s the Germans made sure through the production of high-quality radios that 'tank units are now under guaranteed command and control.' Provision was made for numerous armoured command vehicles, so that even commanders of corps of tens of thousands of troops had mobile, protected HQs close to the action. Guderian would later comment that throughout the Second World War, 'we were at all times superior to our enemies' in radio communications. German tanks all had them, though in a move that might appeal to today's micro-managers, only the commanders could speak, the vehicles of those lower down the food chain just being equipped with receivers.

Despite the efforts made by General Estienne to provide the Renault light tank companies of 1918 with a radio-equipped command vehicle, they had fallen behind. Most of the FCM-36 light tanks encountered in May 1940 had no radio, and the Char B1, comically, was originally only equipped with a set capable of Morse code. These shortcomings, as well as the higher level of inefficiency and paralysis of the French military, had helped the Wehrmacht to run rings around a numerically superior force with thousands of modern vehicles. Effective communications allowed panzer crews to coordinate their actions and manoeuvre around a threat like the Char B1, engaging it from the flank or rear.

The other quantum that Germany's armour designers had exploited was the human one. They realized that good ergonomics helped get the best out of their people. As noted, the commander's position in the Panzer III and IV offered terrific situational awareness. Sitting in the immaculately restored Panzer III at The Tank Museum in Bovington, it's striking how good the all-round visibility is. When spotting the muzzle flash of an enemy anti-tank gun might make the difference between life and death, this advantage really counted for something.

By contrast, the Char B1's commander had been placed in an absurd position by the vehicle's designers. He was the only person in the turret, having to crew the 47mm gun as well as give his crewmates orders, master the tactical situation and navigate across the battlefield. To make matters worse, while he had vision devices in a small raised cupola, it did not have a hatch above it, so he could not easily check things himself by eye if the periscopes let him down.

The question of ways in and out of the vehicle was another one that could make a huge difference to fighting efficiency. In their private thoughts, every crew member knows that their armoured shell can in seconds become the oven in which they will be incinerated. Each of the Panzer IV's crew had their own hatch, if escape became imperative. Likewise, the Panzer III gave the two in the hull as well as the commander overhead escape routes. Both vehicles also had hatches on the turret sides, which loader and gunner could use. It wasn't simply a matter of bailing out: it helped situational awareness, and staving off boredom or motion sickness on long road marches.

The French campaign, and the invasion of Poland that preceded it, had given the *Panzertruppe* an appreciation of their vehicles' strengths and weaknesses. It was readily apparent that the surviving Panzer I and II vehicles, while having some uses in reconnaissance, training or command roles, were already obsolescent. The two newer models would be subject to numerous improvements.

In July 1938, major production had begun with an order for 2,155 Panzer IIIs and 640 Panzer IVs. As the numbers suggest, it was the first of these that was meant to play the principal role in panzer divisions. The Panzer IV, firing high-explosive and smoke shells, was seen as a supporting player.

At one point the HWA considered abandoning production of the Panzer IV and basing the entire fleet on the Panzer III. It was saved by Krupp's engineer Dr Wolfert, who pointed out that the production of his vehicle was gearing up very nicely, whereas in 1939 the other tank was way behind its targets. Thus Wolfert saved his baby, which was just as well for the panzer arm because its bigger size would later mean it had greater growth potential.

By 1940 the technological competition between the warring powers was in full swing. Germany, the Soviet Union, Britain and the United States engaged in a relentless, dynamic arms race, constantly trying to gain the advantage, be it in firepower, protection or mobility. This produced repeated up-gunning, up-armouring and souping-up of engines.

Guderian had before the war lost an argument with the bureaucrats of the HWA over the Panzer III's armament. He had wanted a 50mm gun, they 37mm. After France the decision was swiftly made to up-gun to 50mm, and later to a longer-barrelled, high-velocity weapon of the same calibre. This was to produce a great

1. Tank Corps Mark IVs brought by rail close to Cambrai in November 1917. The fascines, or bundles of wood, atop each one produced a nasty surprise for the enemy.

2. The assembly hall at Fosters, where the pace of production proved inadequate for the War Office and the tank's inventors had to share their blueprints with other manufacturers.

3. In the Tank Corps, names denoted the units; those beginning H, like Hyacinth, were in H Battalion, this one becoming firmly planted in a German trench which it attempted to cross without a fascine to roll over.

4. Few Mark IVs survived the Great War, improved vehicles replaced them, and following the Armistice rapid demobilisation saw most scrapped, but this one is preserved at Bovington.

5. A prototype Renault tank being driven by Albert Stern, a member of the British tank committee who cultivated close relations with the French.

6. A Renault in action in July 1918. By this time large numbers of them were entering service with the French and US armies.

7. Although France sold off many of its Renaults, hundreds remained in service at the time of Germany's May 1940 invasion. Captured ones were pressed into service for such missions as airfield defence and anti-partisan operations.

Although two decades old by the time of the Spanish Civil War, Renault FTs were used by both sides, proving simple to operate and maintain.

9. A column of Panzer IVs captured during the invasion of France. These early models had the 'cigar stub' infantry support gun as their main armament.

10. Not only was the Panzer IV the only tank produced from the start of the Second World War until its end, but it also served on long afterwards, for example with the Syrian army on the Golan Heights during the 1960s.

11. Even in 1944 the Panzer IV with its long 75mm armament remained effective – but by adding spaced armour the Germans hoped to boost protection against the array of anti-armour weapons appearing on the battlefield by then.

12. The additional plate could not save this tank from the bazookas carried by the US infantry, the projectiles of which exploited shaped charge projectiles that could penetrate around 80mm of armour.

13. This Panther was knocked out in Argentan on the 20th of August 1944. While well armed and thickly armoured on its front, the vehicle was large, heavy, poorly protected on its sides and mechanically often unreliable.

deal of 'I told you so', not least from Adolf Hitler himself after the unpleasant surprises received during the invasion of Russia.

In 1940 the Wehrmacht had come up against French armour and measured their own strengths and weaknesses against it. They had also got some insight into the qualities of British tanks. During a counter-attack at Arras, they had been briefly shaken – both by the effective use of a brigade-sized attack against them, and by the difficulty they had in knocking out Matilda II tanks. With its frontal armour up to 78mm thick (compared to the Panzer IV's 30mm), it could take a lot of punishment, even if it had been poorly armed for its mission of infantry support.

The Arras counter-attack had broken into the rear of the 7th Panzer Division, which was, during the invasion of France, under the command of Major General Erwin Rommel. Aged 48 at the time, he had won his spurs as an infantry commander during the Great War and in 1937, as an instructor at the war college, published an influential book on tactics.

Rommel had come to the Führer's notice when commanding his escort a couple of years later. Under Hitler's spell, and flirting with Nazi ideology, this posting opened the way to the plum assignment of commanding a panzer division, even though he had no particular expertise in mechanized warfare.

Under him, the 7th would advance 150 miles during the French campaign, and Rommel would complete a rapid apprenticeship in tank warfare. Despite briefly being rattled by that British counter-stroke at Arras, his division had performed very well during this campaign, and early in 1941 he found himself assigned the mission of taking a small corps of armoured and motorized forces to bolster the Axis forces in North Africa. In this campaign he would repeatedly do battle with British armoured forces.

On the 2nd of April, advancing elements of Rommel's Deutsches Afrika Korps or DAK came into contact with the British for the first time, during a swift encounter on the Libyan coastline. The men of 5th Royal Tank Regiment, crewing a squadron of A13 tanks, traded fire with a mixed group of Panzer IVs and Panzer

IIIs, the latter being armed with the 50mm gun introduced following experiences in France.

British tanks knocked out three panzers and lost five of their own in the action. Six A13s escaped the initial scrap but most broke down as the British fell back eastwards to Tobruk, only two of them making it to the fortress. The salient message of these early encounters was that British 'cruiser' tanks like the A13 and its relatives the A10 and A12 deployed in the Middle East were outclassed because of their thin armour and poor mechanical reliability.

In the headlong advance eastwards that ensued, the DAK's panzers, once new air filters were fitted, racked up what British 'tankies' would have considered astonishing mileages. This was in large measure due to the exacting engineering standards to which vehicles like the Panzer IV had been built.

The other great lesson of these early actions in the desert was that the possibilities for the successful use of armour depended greatly on the terrain. North Africa provided a clear battlefield where armour could use its weapons to maximum effective range because interruptions to sight lines like villages, trees or hills were notable by their absence. In such an environment, questions such as the quality of tank gunnery – weapon, optics, crew and training all combined – were critical. Similarly, the absence of river lines and scarcity of other natural obstacles allowed for mobility to be exploited fully too.

Rommel wrote of the desert: 'It was the only theatre where the principles of motorized and tank warfare . . . could be applied to the full . . . where the pure tank battle between major formations was fought.' A matter of months after those initial desert encounters the Wehrmacht would find itself in an environment almost as well suited to the employment of panzer formations: the plains of Eastern Europe. If the need for a longer version of the Panzer III's 50mm gun, with greater penetrative power, was not already clear early in 1941, the invasion of the Soviet Union would soon underline that point.

Late in the spring of 1941, reflecting on the triumph in France and the impressive advances of the Afrika Korps, as well as what might lie ahead for their forces, Hitler and the German general staff had not yet got to grips with another key question. There was no doubt that German industry could produce quality, but what about quantity?

Having begun rearmament in earnest in 1935 and settled on a doctrine that put tanks centre-stage only in 1938, Germany faced a real struggle gearing up production fast enough. Britain, having fallen to a low of just 16 tanks built in 1930, was by 1938 up to 1,087. France had begun to ramp up far earlier. The unquestioned champion in numeric terms was the Soviet Union, which in 1938 produced 2,270 tanks.

Germany's issues were all the more acute because it formed nine additional panzer divisions during the year between the conclusion of its French campaign and Operation Barbarossa, the invasion of the USSR in June 1941. Yet for a regime bent on conquering all-comers, that had placed enormous faith in the panzer force, Hitler's Germany was remarkably slow at making them.

In 1938, monthly production of the Panzer IV usually barely reached double figures. By 1940 it had been increased, but still just 260 of the tanks were made that year. Keep in mind that 70 had been lost in Poland and dozens more in France. That average of 20 Panzer IVs a month compared to a target of 50 that had been given to Krupp.

By way of comparison, production during 1940 of the two main Soviet types (T-26 and BT-7) that the Germans would face when they did invade was more than 2,400. As in France, the German high command's understanding of what was going on in the Russian tank industry was faulty to say the least. To the slowness of Germany to formulate its doctrine, and produce the machines required to make it happen, can be added poor technical intelligence about what its rivals were up to.

Naturally, having taken France and entered into a global war, it was evident that things could not remain as they were. Boosting Panzer IV output required new plants. So, in July 1941, bus-makers

VOMAG (standing for Vogtländische Maschinenfabrik AG) fired up production at Plauen, a purpose-built plant in south-east Germany. A few months later the Nibelungenwerk in Austria also started making them. With three factories on-line, Panzer IV monthly production reached 61 in December 1941.

Of course, this wasn't the only tank available to the Wehrmacht, and Panzer III figures were higher during 1941. However, with the new plants on-line, the Panzer IV outstripped them in 1942. But in order to equip their new panzer divisions that wasn't enough, so the Germans resorted to expedients like using the Panzer 38(t), an inter-war Czech design. Having annexed Czechoslovakia along with its considerable armaments industry, Germany ordered 1,400 of these for its own forces. The Panzer 38(t) was poorly armoured and had the same 37mm gun shown to be inadequate in France. Nevertheless, for a time it became the mainstay of entire panzer divisions.

This tendency to exploit the industrial capacity of its vanquished neighbours was applied even more widely to wheeled vehicles. While the Wehrmacht considered French tanks rather second-rate, they happily acquired tens of thousands of motor vehicles from military stocks and new production. Louis Renault, one of the heroes of our last chapter, earned lasting opprobrium among his people for the alacrity with which he turned out vehicles for the German occupier.

There are various explanations as to why, having embarked on a world war, the Nazi leadership did not make better preparations for the mobilization of its national industries, as well as those of the broad swathes of Europe that were conquered: it had been given insufficient political focus; Hitler either wasn't interested early on, or meddled too much later; and the panzer arm in general did not rank highly enough in the order of priorities.

Certainly, the panzer generals found it inexplicable that huge amounts of steel still went to naval construction once the German fleet was effectively bottled up, or that production of railway locomotives and other civilian transport continued at such a high pace.

Some historians have also pointed to exquisite engineering standards and complex supply chains as a reason why panzer production remained so low. Museum staff who have restored Panzer IIIs and IVs compare the build standard to a Swiss clock rather than the brute ironmongery of Soviet designs.

To all those reasons must be added the blind conceitedness of National Socialist ideology and the hubris of a general staff that had just crushed the forces of its historic Gallic enemy. Retrospectively we should be thankful for it.

For in the early summer of 1941, as Hitler prepared to launch his armies into the Soviet Union, some of his generals had forebodings, while others assumed that the innate superiority of German arms, of the blitzkrieg system, would carry the day. But as for the panzers that played such a central part in that way of battle, there was an assumption that they were unbeatable. 'We believed', Guderian would later write, 'that at the beginning of the new war we could reckon on our tanks being technically better than all known Russian types.' How wrong they were.

The flood unleashed on the 22nd of June 1941 ranks as the greatest onslaught in military history. Germany and its allies brought 3.8 million soldiers to the fight, as well as 3,000 tanks and other armoured vehicles. The Red Army had over 2 million troops, but its impressive industrial efforts ensured that it had more than 10,000 tanks ready in the western part of the USSR, as well as thousands more undergoing light repair or stationed further east.

Operation Barbarossa, in its early weeks at least, is remembered by its participants as an endless advance across the dusty plains of Byelorussia, Ukraine and the Baltic republics. A series of thrusts created the possibility for great encirclements that yielded 3 million prisoners before 1941 was out, claimed the lives of a further 1 million Red Army soldiers and harvested thousands of Soviet tanks.

By the end of 1941, the effective Russian tank strength in the west of the USSR was around 3,300 vehicles, meaning two-thirds of their fleet on the 22nd of June had been written off by the Nazis. We will see why this did not prove decisive, but there were many on both sides during this early phase of the struggle who assumed a German victory was eminently possible.

During the first three months of this vast campaign, the Panzer IVs in the 4th Panzer Division motored an average of over 6,800 miles each. Of course, there were many breakdowns, but the fact that so many of the vehicles were still going after so many miles was a testament to the build quality of Krupp-Grusonwerke. By way of comparison, one British armoured regiment that moved nearly 250 miles through the Libyan desert had 26 of its 53 A10 and A13 cruiser tanks break down along the way.

Reliability was vital, but it wasn't everything. A few months later, amid scenes of jubilation in the Wehrmacht following rapid conquest, legions of prisoners, and the presentation of welcoming bread and salt by subjects of the Soviet empire who regarded Barbarossa as liberation, the tank crews had had a very unpleasant surprise.

Less than ten hours after the invasion began, crewmen of the 7th Panzer Division encountered an enemy they knew nothing about – a brand-new Soviet tank. Indeed, they didn't even initially know its name: the T-34. Crossing a bridge in the Lithuanian city of Alytus, a Panzer 38(t) was swiftly knocked out. The Russian tank's 76.2mm gun completely outclassed the Czech vehicle's 37mm one in range and penetrative power.

The experience of being in a 38(t) knocked out by a T-34 (three weeks later) was recorded by crewman Otto Carius: 'It happened like greased lightning. A hit against our tank, a metallic crack, the scream of a comrade, and that was all there was! A large piece of armour plating had been penetrated next to the radio operator's seat. Nobody had to tell us to get out. Not until I had run my hand across my face while crawling in the ditch did I discover that they had also got me.' Carius had learned a brutal lesson in the power of superior gunnery, and he will return later in our story.

In this and subsequent discussions about tank duels, the key to victory was to beat the armour by kinetic energy – the force contained in an armour-piercing shot. This is usually expressed as the mass of that metal penetrator multiplied by its velocity squared over two. The Soviet 76.2mm gun that defeated Carius' 38(t) fired a shot weighing 6.6kg that left the barrel at 680 metres per second. The shell fired by the Panzer IV's short 75mm gun had a similar mass, 6.8kg, but travelled markedly slower. That difference in speed, squared, made the 'cigar stub' German gun inferior as a tank-killer. As for the 38(t), its 37mm shot flew very quickly, but its mass of just 0.85kg also put it at a big disadvantage.

At Alytus in June 1941, a battalion of 44 T-34s held up the 7th Panzer Division for hours. Eventually, the divisional commander managed to call up Luftwaffe airstrikes and the Red Army tanks began running low on ammunition, yielding their position.

These new tanks formed only about 10 per cent of the tank inventory facing the invasion, and in any case the ability of a few successful, well-equipped units to hold back the general tide was very limited. If troops were falling back to your left and right, after all, you would soon face the danger of being surrounded.

In the case of the action at Alytus, the defending Russian tank division was outmanoeuvred, losing 73 tanks to the 7th Panzer Division's 11. A force that employed all arms skilfully could compensate to some measure for the superiority of the Soviet tank by using air attack, heavy artillery firing at close range or, when even nearer, infantry assault using explosive charges.

There were other unwelcome discoveries for the Panzertruppen: the 44-tonne KV-1 heavy tank, equipped with the same gun as the T-34, and its ugly brother, the KV-2 with an enormous 152mm howitzer. In places there was consternation, panic even, when these new Soviet tanks broke through, and German infantry found their standard anti-tank guns, the 37mm Pak 36, having almost no effect on them.

Lieutenant Carl-Ulrich Dirks, manning a 'Pak front' or unit of 37mm anti-tank guns deployed with interlocking fields of fire, described the shock of a T-34 attack: 'we opened fire at 100m . . . and when we hit the tank the tracer was going up into the sky and we thought we had been given faulty ammunition, we couldn't understand what was happening.' The Red Army tanks rumbled through their position, Dirks recalling: 'we were in despair because we were just helpless.' It took air attacks to turn the tide.

A couple of days after the invasion began, another German division in the Baltic republics was attacked by a Russian tank force which sent forward its armour in waves. The initial attack featured the two types – T-26s (a copy of a British design) and BT-7s (light, very quick) – that made up the bulk of the Soviet fleet. The Germans had contempt for these vehicles and swiftly knocked them out that day.

But the new tanks shocked them profoundly. During the attack of the 24th of June, T-34s, KV-1s and KV-2s formed a follow-on echelon, overrunning a German motorcycle battalion and some artillery positions. In a couple of celebrated cases the Soviet heavy tank crews fought on until the bitter end against large numbers of Germans – earning them grudging respect from the invader and posthumous glory from Kremlin propagandists.

After fighting against T-34s in the battle for Oryol, in October 1941, Major General Willibald von Langermann reported back to headquarters: 'the Russian tanks usually formed in a half-circle, open fire with their 76.2mm guns on our panzers already at a range of 1,000m and deliver enormous penetration energy with high accuracy.'

The Panzer IIIs could only penetrate the Russian tank's sloping armour on the flanks and at very close range. Panzer IVs were also comprehensively outgunned. The general went on: 'many times our panzers were split open or the complete commander's cupola of the Panzer III and IV flew off from one frontal hit.' Inevitably, Langermann wrote, the men understood that they had been overmatched, with morale affected accordingly: 'the panzer crews know they can be knocked out at long range by enemy tanks and that they can achieve only a very minimal effect on enemy tanks.'

By November 1941 an intensive discussion had started about how, as Guderian put it, 'to help us regain our technical supremacy over the Russians'. This would eventually produce a drive to copy the T-34, but everyone knew that creating a new panzer would take time. In the meantime, urgent measures had to be taken.

It was decided to fit Panzer IIIs with a longer, high-velocity 50mm gun, a change that got underway that December. The army ordnance office's previous disinclination to use such a weapon became a rod with which to beat them, armaments minister Albert Speer writing: 'Hitler was triumphant because he could then point out that he had earlier demanded the kind of long-barrelled gun [the T-34] had.' It was this tank gun saga that thereafter prompted the Führer, convinced of his wisdom in such matters, to micro-manage armoured vehicle designs.

The second development – actually a critical milestone in the tank story – was the introduction of a new shell for the Panzer IV's 75mm gun. Until this point, penetrating armour was a brute-force business of firing a solid metal shot out of a gun with such force that it would smash its way through metal. When the enemy fielded thicker armour, it meant increasing the size and speed of your anti-tank round, imparting to it more kinetic energy, so it would still break through.

Germany's new shell used a very smart trick of physics to convert the explosion when it hit its target into something of great penetrative power. Experts had long understood that if explosives were shaped to create a cavity, usually an inverted cone, the resulting blast could be focused to a point just in front of the device. If you lined the conic interior with metal, the blast would, in a fraction of a second, invert that liner into a penetrative bolt flying at enormous speed.

This 'shaped charge' technology had advanced in several countries, including Germany, to the point that demolition devices designed to cut deep into concrete bunkers were used during the 1940 campaign. Now reducing it to the 75mm diameter of a shell, German experts realized that the resulting munition, though it matched the existing one in terms of speed and trajectory, could pierce dense armour. Tests showed it would cut through up to 75mm of steel.

The advent of this new type of explosive charge was a turning point in the history of the tank – indeed, it was so important that it can be argued that the armoured vehicle has never quite recovered from it. The shaped charge, later dubbed High Explosive Anti-Tank or 'HEAT' ammunition by Western countries, gave all manner of vehicles, and later even hand-held rocket launchers, the power to overmatch thick steel armour. It offered an alternative to the business of trying to create ever more powerful guns that could impart to a steel shot enough kinetic energy to penetrate the thickest armour, though to this date it has not entirely replaced that technology.

Shaped charge shells were sent to units in Russia at the start of 1942 and soon proved their worth. One battalion reported successful engagements against T-34s and KV-1s, saying of the new shell: 'it simply sticks to the surface and then penetrates. The process resembles a mechanic using a welding torch.' The new shaped charge round meant that, while a Panzer IV was still best at firing explosive shells in support of infantry, it stood a far better chance of surviving an encounter with one of the latest Soviet tanks.

Change of a more fundamental kind was also on its way. A new 75mm anti-tank gun was under development that offered a big increase in penetrative power. This weapon, called the Pak 40 when mounted on a wheeled carriage, fired shells at 790 metres per second (as opposed to the 450m/s for high-explosive rounds hitherto used by the Panzer IV) and could penetrate more than 80mm of armour at 1,000 metres. A tungsten-based armour-piercing round offered even more impressive performance, though shortages of this metal prevented its widespread use.

The key point though – and it was already well understood in research establishments from Magdeburg to Maryland – was that by making the barrel longer, more of the energy from the blast in the weapon's breech got transferred to the shell. Switching out the Panzer IV's short barrel for a longer 75mm made a dramatic difference. Tanks given the new gun were soon being referred to as Panzer IV 'longs' in Wehrmacht paperwork.

These technical measurements translated into something much more fundamental: the previous classification, adapted from the British, of cruiser and infantry tanks became meaningless. The Panzer IV had been given a weapon that made it a more effective tank-hunter than the Panzer III (even with its long 50mm gun), but it could also still fire high explosive or machine guns in support of the infantry. When continued armour and engine improvements were taken into account too, it would become the panzer divisions' mainstay.

From then on, for around 18 months, the Panzer IV entered a purple patch, a moment when the design had matured to make it one of the best tanks on the battlefield. True, the T-34 maintained an edge in certain respects, but, with its impressive firepower and excellent ergonomics, the German vehicle was hard to beat.

Early reports from the Afrika Korps, where the long-barrelled vehicles had also been deployed, noted: 'the gun proved superior to all hitherto used weapons during first engagements.' The only downsides, from the operator perspective, was that the more powerful weapon produced more flash and dust, thus attracting British fire towards Panzer IV longs, which they regarded as a priority target.

By this time also, with production stepping up at three factories, German industry was at last able to produce significantly more of them. By the latter part of 1943, more than 300 were being made each month.

And what of that sibling nudged off centre-stage, the Panzer III? Large numbers still remained in service throughout 1942. Equipped with the long 50mm gun they were very effective against the British in North Africa, and could hold their own against many of the Russian models too. But the Panzer III's turret was not large enough to accept the long 75mm weapon.

That tank's chassis had already been used as the basis for the *Sturmgeschütz* III, a turretless vehicle with the same infantry support howitzer as the early Panzer IV. Equipped thereafter with the long 75mm gun, and up-armoured (with 80mm at the front), the StuG III Ausf. F, as it was abbreviated, became a highly effective tank-hunter from the late spring of 1942 onwards. This new vehicle was perfect for ambushing. It had a lower profile than a tank, was well protected and packed a powerful punch. Indeed, the StuG would become the vehicle used by many of Germany's 'panzer aces'.

By switching production from the Panzer III tank to the StuG variant, Germany reaped certain industrial rewards too, because the tank-hunter didn't require the large bearings or servos, among other items, needed by the turreted vehicle. It also became possible, a few months after production of the F variant StuG began, to fit a 75mm barrel longer than that used by the Panzer IV, further improving its ability to penetrate enemy armour. By August 1943 production of the Panzer III tank had stopped, but that of the StuG continued and would, with more than 8,400 made, eventually exceed that of the Panzer IV, becoming the most numerous German armoured fighting vehicle of the war.

The success of the StuG III touched off a sort of 'what is a tank?' debate, since evidently it confounded those who believed a turret was a defining characteristic. After all, when supplies of replacement tanks proved inadequate, the vehicle ended up equipping many battalions in armoured divisions, and it was crewed by the same Panzertruppen who might operate Panzer IVs.

Thus, the success of this new tank-hunter caused some unease among panzer generals. Guderian believed that Hitler had bought into a number of heresies. The Führer was convinced that the shaped charge shell, once adopted by Germany's enemies, would limit the role of the tank, traditionally defined. Hitler also felt the armaments ministry should raise the priority given to tracked assault guns and tank-hunters.

Heinz Guderian, who re-emerged as a central character in this following his appointment in March 1943 as inspector general of panzer troops, did everything he could to continue large-scale tank production and stymie the shift towards the turretless designs. In his notes of arguments to be used on the Führer, the general wrote: 'the assault gun lacks the decisive, lightning-fast ability to operate in all directions, and the ability to fight without support of infantry.'

Guderian fought repeated battles to maintain production of the Panzer IV, making the argument to Hitler that 'a conversion of PzKpfw [Panzerkampfwagen] IV production to assault guns would mean turning a versatile weapon into a one-dimensional one.' Even in 1943 when significant numbers of what he called the 'problem child', the new Panther tank, started to appear, the general argued that the vehicle was immature and the Panzer IV the Wehrmacht's only sure bet.

While these debates went on in Berlin, those engaged in the ugly reality of Eastern Front fighting were supposed to rely on superior skills as well as technological refinements to maintain their edge over the Red Army. In combat during March 1943, the Grossdeutschland Division pitted a force of 79 tanks and 19 StuGs against its Russian foe. In just under a fortnight it claimed a 20:1 kill ratio, saying it had knocked out 269 Russian tanks, with the Panzer IV longs racking up 188 and the StuGs 41.

Both sides inevitably exaggerated their kills, but there would be a number of well-documented cases where the Germans destroyed grossly unequal numbers of tanks in their fights with the Russians. By the time of those engagements catalogued by the Grossdeutschland Division, Germany's top-tier predator, the Panzer VI or Tiger, had also appeared on the battlefield. However, numbers were small and the Panzer IV long had evidently been a very impressive performer in those engagements.

The precise exchange rate between those tanks and T-34s, or the odds of victory in a one-to-one engagement between them, were largely beside the point. The vastness of Soviet resources, and the support of its Western allies, allowed Joseph Stalin to accept hideous losses.

It wasn't just about numbers either, because in the T-34 the Russians had produced a great design and were doing everything possible to ensure its manufacture at astounding levels.

T-34

Entered service: 1940
Number produced: 35,000 (of 76.2mm gun variant; a further 23,000 T-34/85)
Weight: 29 tonnes
Crew: 4
Main armament: 76.2mm gun
Cost: R 130,000–250,000 ($24,500–47,000 at wartime exchange rates, equating to £8,000–11,750, the lower figure equating to £339,000 in 2024, though really any prices in the 1940s Soviet command economy need to be taken with the largest pinch of Siberian salt)

T-34

The 12th of July 1943 began confidently enough for men of the SS Leibstandarte, pushing forward in their tanks towards the Soviet town of Prokhorovka, close to today's frontier between Russia and Ukraine. Obersturmführer von Ribbentrop sat in the cupola of his Panzer IV scanning the open farmland, great rolling fields belonging to the *kolkhozes*, the Soviet collectives. The rank was equivalent to lieutenant, and he was in charge of a company of a dozen or so panzers.

It was around 8 a.m., and his orders were to push north-north-east as part of the ongoing German offensive, codenamed Citadel, by which Hitler and the high command hoped to regain the initiative. Ribbentrop had no idea that a Soviet counter-stroke – one of the biggest tank attacks in history, in fact – was about to meet him, coming in the opposite direction.

He had just crested one of the low ridges that punctuate the steppe landscape, when ahead and to his left an altogether unexpected panorama opened up: 'As we moved down the forward slope, we spotted the first T-34s. They were attempting to outflank us from the left. We halted on the slope and opened fire, hitting several of the enemy . . . for a good gunner 800 metres was the ideal range. As we waited to see if further enemy tanks were going to appear, I looked all around as was my habit. What I saw left me speechless. From the shallow rise about 150 to 200 metres in front of me appeared fifteen, then thirty, then forty tanks. Finally, there were too many to count. T-34s were rolling toward us at high speed, carrying mounted infantry.'

Lieutenant General Pavel Rotmistrov had just unleashed one of the Red Army's biggest tank formations, a key part of Stalin's 'supreme reserve', in a bold attempt to crush the 'Fritzes' before their offensive could gain any more ground. Bespectacled and moustachioed, the general was a smart political operator who had survived the purges prior to rising to high command. Having graduated from the Frunze Military Academy and been chief of staff of a mechanized corps he had served the necessary time, learning his trade before being given a critically important job.

His 5th Guards Tank Army fielded more than 900 tanks that day, as well as dozens of self-propelled guns, and was central to the bigger Soviet plan. Crouching on many of the armoured vehicles were paratroopers and other infantrymen, 'tank riders' who would be carried forward, leaping off when the moment came to take an enemy position.

'I'm surprised by how quickly our tanks and the hostile tanks are closing the distance to each other,' Rotmistrov later wrote, recounting how he watched the advance through his binoculars: 'two enormous avalanches of tanks are moving towards a collision.'

Ribbentrop knew that his small group of tanks would soon be overwhelmed: 'I said to myself: "It's all over now!"' His driver, hearing curses, tried to bail out of the vehicle, but was pulled back in. Meanwhile the turret crew began fighting frantically. 'We had no time to take up defensive positions. All we could do was fire. From this range every round was a hit.' One of his Panzer IVs a short distance away was engulfed in flames; Ribbentrop spotted the commander bail out, but never again after that.

Directing the retreat of his beleaguered group, the SS officer knew he would have to get back towards German positions with his vehicles blended in with the enemy's – would they be knocked out by their own side? He also knew that an anti-tank ditch lay behind them, a Soviet obstacle the panzers had themselves just crossed a little earlier that morning.

It seemed no more than minutes after the engagement had started, with burning T-34s dotting the fields around them as dozens of others moved past, that Ribbentrop realized he was about to run out of armour-piercing ammunition for his 75mm gun. The radio operator down in the hull started emptying the storage bins below, passing rounds up to those in the turret.

Wilhelm Roes, another member of the same 1st SS Panzer Regiment, watched Ribbentrop's company moving back towards the anti-tank ditch. It 'had been almost completely overrun and in retreating, it was drawing Russian tanks after it in close pursuit.' The surviving Panzer IVs made it over the obstacle, then Roes watched as the T-34s, speeding over the edge, dropped a couple of metres into the enormous trench. He shuddered at the crunch of flesh and bone that must be going on inside the Russian vehicles. Then, to his amazement, he saw some of the T-34s climbing out the other side: 'Our tanks would not have made it through such a scenario – their tracks wouldn't have held up to such a shock, but the T-34s kept moving.' History does not record how many of the tank riders managed to hold on.

Prokhorovka was proving to be the climax of Citadel, or as many would subsequently refer to it, the Battle of Kursk. Everything about the scale of the German breakthrough attempt, launched on the 5th of July 1943, and the Soviet counter-attack was enormous: in all, 4 million troops were involved, fielding 69,000 artillery pieces, thousands of combat aircraft and 13,000 armoured fighting vehicles.

Months of planning had gone into Citadel, giving the Red Army plenty of time to build successive defence belts. The Germans aimed with two great pincer attacks to lop off the bulge of Soviet forces occupying the Kursk salient. But the Red Army had planted nearly 1 million mines, an average of 1,600 per kilometre of front, and brought in several brigades of anti-tank guns, as well as digging ditches that were supposedly impassable to any vehicle.

It is remembered as a titanic tank duel, and indeed it was, even if the crews were just a small part of each side's order of battle. Of these, the T-34 was by far the most numerous on the Russian side. Among their opponents the Panzer IV formed the mainstay, though there were hundreds of Panzer IIIs with long 50mm guns present too, as well as some interesting novelties.

The Tiger tank, which already had a gruesome reputation among Allied crews, was there, and indeed was everywhere in the reports of Soviet commanders, though just 117 were present across the entire tract of southern Russia where the offensive took place. German factories had also managed to get a couple of hundred of the new Panther tanks into action for Citadel, and nearly 100 Ferdinand heavy tank-destroyers. However, both Panther and Ferdinand delivered a lacklustre performance in this battle.

In the north, one arm of Hitler's attempt to sever the Kursk salient had soon run into difficulties. But the forces coming up from the south, spearheaded by II SS Panzer Corps, made better progress. The Leibstandarte division (properly the 1st Leibstandarte SS Adolf Hitler Division) was that morning advancing in the centre, with the 2nd SS Panzer Division (Das Reich) to its right, a little further back, and the 3rd SS Panzer Division (the Totenkopf or 'Death's Head') to its left. The attack by Rotmistrov's 5th Guards Tank Army at Prokhorovka fell mainly on the Leibstandarte and Das Reich.

Red Army crews had in many places faced this offensive with trepidation. Lieutenant Vasiliy Krysov described witnessing panzers advancing towards them early on during Citadel: 'We now could clearly see through the periscope and gunsights as the Tigers advanced, slightly weaving as they prowled through the wheat field, their menacing gun muzzles swinging back and forth as they scanned our positions for targets.' Krysov, who commanded a self-propelled 122mm howitzer, let fly one round without effect, but then managed to dislodge a track with a second. Some accounts speak of the tank men crossing themselves, falling back on religion despite years of Soviet atheism.

That July morning, Lieutenant Vasiliy Bryukhov was commanding a T-34 that attacked the flank of the Leibstandarte near Prokhorovka. His vivid description, recorded many years after the battle, may not have the advantage of immediacy but it certainly painted a more honest picture than most of those released soon afterwards: 'The distance between the tanks was below 100 metres – it was impossible to manoeuvre a tank, one could just jerk it back and forth a bit. It wasn't a battle, it was a slaughterhouse of tanks. We crawled back and forth and fired. Everything was burning. An indescribable stench hung in the air over the battlefield. Everything was enveloped in smoke, dust and fire, so it looked as if it was twilight.'

Captain Vakulenko, sent forward with a column of T-34s in an attempt to extricate the unit that had charged headlong into the anti-tank ditch that morning, recorded: 'The fascists concentrated a storm of fire on our tanks. Laying down a broken track, one tank came to a stop; another T-34 exploded. Over the radio I gave the command, "To everyone, to everyone, to everyone! Smoke!" A thick shroud of smoke enveloped the tanks.'

On the other flank of that SS division, dozens of Russian tanks managed to exploit the seam between it and the Totenkopf to its left. Two fingers of the German armoured thrust had parted a little, and the 18th Tank Corps (part of Rotmistrov's army) tried to exploit it. In places the T-34s managed to get up onto higher ground, encountering German infantry and artillery positions. In this desperate close-range fight, tanks crushed guns and their crews, and once or twice may even have rammed each other. German artillerymen found themselves firing over open sights at T-34s, blasting them with high explosive.

Attempting to check the advance at the seam of the two SS divisions, a platoon of tanks was ordered forward under Michael Wittmann, a 29-year-old Bavarian lieutenant. He had served an apprenticeship as a commander of StuG assault guns and Panzer IIIs, having graduated to a Tiger just three months before Citadel. Wittmann became a panzer ace, one of those figures beloved by Nazi propagandists who found the parallels between the armour-plated commanders of 1943, and Teutonic knights heading east in medieval times, impossible to resist.

Wittmann led his platoon of four Tigers forward to block the emerging penetration on his division's flank. It isn't clear exactly how many Soviet tanks they knocked out, but it was dozens, perhaps as many as 50. Having brought his platoon onto a forward slope in order to do this, Wittmann's vehicles were soon being struck by Soviet shells. One of his Tigers, immobilized, was left stranded and lost.

Back at the headquarters of the 5th Guards Tank Army, it was becoming evident as the day wore on that, far from unleashing a tide that would sweep all before them, Rotmistrov was struggling to hold on to the meagre gains that had been made. One of his staff officers overheard the general shouting at one of his subordinates over the radio: 'Tell him: Not a step back! Things are difficult for everyone, everyone is suffering losses.'

As the fighting ebbed at the end of the day, it was apparent that the 5th Guards Tank Army had suffered a setback on an epic scale. Faced with the awful truth, Rotmistrov, a survivor if ever there was one, clearly understood what might happen to him. He resorted to the option Soviet bureaucrats knew only too well: he lied, claiming to have destroyed 298 German tanks, 55 of them Tigers.

In subsequent telling, the general's staff insisted that it had been an unequal fight, a sort of T-34 David versus the Tiger Goliath, and no matter that the great majority of panzers encountered were IIIs and IVs. The Russian report argued: 'The gun on the T-34 tank was unable to destroy the enemy's medium and heavy tanks at a range of 600 metres and greater, while the gun on the enemy vehicles could damage our tanks even out to a range of 1,200–1,500 metres.' Staff in Rotmistrov's army also pointed to the hurried way their attack had been launched, leaving too little time for 'battlefield preparation' by air and artillery attack in order to neutralize German anti-tank defences.

Stalin most likely knew that he was being fed lies and excuses. He reportedly asked Rotmistrov soon after the battle, 'What have you done to your magnificent armoured army?' There was talk of a special tribunal being set up to investigate that exact question. Rotmistrov would not have been human if he did not contemplate the possibility of disgrace, dispatch to the Gulag, or death.

Publicly, though, Soviet propaganda had to trumpet it as a great success. At the museum built to commemorate the feat, a vast sculpture depicts a fictitious scene of T-34s ramming a Tiger. It has fallen to subsequent historians like the Russian Valeriy Zamulin and Briton Ben Wheatley to puncture the 'myth of Prokhorovka'. In the latter case, the discovery in an archival backwater of later armoured vehicle returns for II Panzer Corps and cross-checking with aerial reconnaissance photographs taken soon after the 12th of July allowed the truth to be established.

Far from losing 298 tanks on the 12th of July, Wheatley established that the SS divisions engaged lost a maximum of 14. And the 55 Tigers claimed by the 5th Guards Tank Army? We now know that just four of these tanks were engaged, Wittmann's platoon, suffering a single loss.

If the picture on the German side has now been set right, what about the losses of Rotmistrov's tank army? Scrutiny of armoured vehicle returns for the 5th Guards Tank Army, kept secret at the time (understandably enough), indicates that it lost 334 armoured fighting vehicles to the German 14. This Soviet toll was made up of 222 T-34s, 89 T-70s (a light tank usually used for scouting), 12 Churchills (supplied the previous year by Britain under the Lend-Lease scheme) and 11 self-propelled guns.

As so often with tank warfare, the question of who retained possession of the battlefield as night fell was to prove critical at Prokhorovka. So many vehicles were abandoned in battle after becoming immobilized that recovery of such machines could change the calculus of victory considerably. The Germans were in possession of most of this fighting ground for a few days afterwards, sending demolition squads to blow up any abandoned vehicles that might be recovered, ensuring those 334 Soviet losses were permanent.

The battle was highly unusual, not just in scale, but because in many places large numbers of tanks slugged it out against each other. So many tanks were funnelled into a relatively small tract of countryside that other weapons were relegated to secondary importance. It was a real test of armoured technology and crew quality. Anti-tank guns, mines and the infamous ditch may have accounted for many of the 300-plus Soviet losses, but panzers did most of the destruction.

The enormous margin of Red Army losses must, to an extent, undermine the arguments of those who extol the T-34's design – for in most places it was up against Panzer IIIs and IVs. Their ability to kill T-34s had been transformed since the early days of Barbarossa by the fitting of long 50mm and 75mm guns, but other factors were relevant too, from superior crews to elements of the Soviet design such as poor ergonomics. It was one thing having thick, sloped armour and a decent gun, but if the crew struggled to see anything when closed down they were at a distinct disadvantage.

Yet despite the dramatic disparity in the Wehrmacht's favour, Prokhorovka did mark a culmination point for Operation Citadel. Just ten days after that clash, Hitler called time on the operation. He did it for a number of reasons, most importantly that the original aim of lopping off the Kursk salient was clearly unattainable, with the Red Army still showing a great deal of fight, launching large-scale attacks elsewhere. It is also true that Hitler realized, following the invasion of Sicily, that a new front was opening up, and ordered significant reinforcements, including some parts of II SS Panzer Corps, to redeploy there.

Rotmistrov later bragged in his memoir that after the 'bloodbath' unleashed by his army, 'the fascist command was compelled to cancel the offensive.' Given what we know of his untrustworthiness, it is as well to seek views from the other side. General Heinz Guderian wrote, a few years after the Second World War: 'By the failure of Citadel we had suffered a decisive defeat . . . From now on the enemy was in undisputed possession of the initiative.'

So the question is how on earth, when the battle of the 12th of July – arguably the biggest tank fight in history – was such a disaster for the Red Army, did it mark a turning point in their favour? And given that our subject is the T-34, what role did that vehicle play in turning the tide?

On the 7th of August, just a few weeks after Prokhorovka, the 5th Guards Tank Army launched a new offensive near Belgorod. It had, according to Ben Wheatley, been 'fully re-equipped'. Soviet industry and the Red Army training system had brought Rotmistrov's battered command back to more than 660 tanks and crews. The army commander was not wrong when he stated that the T-34, armed with its 76.2mm gun at least, was losing the edge it had enjoyed during the early days of the German invasion more than two years before. But it remained the mainstay of his army, and of Soviet tank forces more widely. The story of the T-34 may have been one, early on, of the ingenuity of Soviet design, but it became the perfect example of the country's astonishing resilience and the success of its mobilization for total war.

At the time that the T-34 was born, the USSR tank industry was still very young, though it was growing at a giddying pace. During the early years following the October Revolution of 1917 and the triumph of the Bolshevik Party, ideas about tank development were entirely based on foreign designs. The Renault FT became the Red Army starter tank, as it was for so many other countries. The Russian authorities were keen to establish their own production, the 'Russkiy Reno' factory being built in Gorky (Nizhny Novgorod, around 250 miles east of Moscow).

Of the various designs that followed, it was the T-26 that ended up being produced in the largest numbers, but it was simply a licence-produced copy of the Vickers 6-tonner, a British private venture design that had appeared in 1928, entering Red Army service three years later. This period – the late 1920s and early 1930s – coincided with the setting up of big factories using imported American machinery and assembly line know-how, an investment that would later prove critical in outproducing the Germans.

Design-wise, things got more interesting a couple of years later when Soviet engineers started adapting an American technology (developed by engineer J. Walter Christie) to vehicles of their own. Christie's genius was to combine a new suspension system and drivetrain to allow vehicles to move very fast. The *Bystrokhodnyy Tank* (the BT series), literally 'fast-moving tank', could be used without tracks on-road (at speeds up to 43mph) and, once the caterpillars were fitted, Christie suspension gave the wheels markedly more travel than other designs, enabling them to soak up bumps off-road, delivering impressive cross-country performance too.

Every engineering solution has its compromises, though. The ability to drive the road wheels (those in contact with the road, unpowered in almost all other tank designs) took up space, adding complexity and weight. Running without tracks meant the front road wheels had to be steerable, so ditto for space, complexity and weight. The trailing arms used to spring the wheels came at a cost too. Thousands of the resulting tanks, notably the BT-5 and BT-7, were made. They were highly mobile, and well-armed too (once a 45mm gun was adopted), with protection inevitably coming third as a priority.

The Soviet Union sent hundreds of T-26s and BT-5s to Spain following the outbreak of the Civil War there in 1936. Although the armour was ostensibly part of the Republican army, a large number of Red Army specialists went with them to maintain and often crew these vehicles in battle. And it was their reflections on this experience of battle that prompted their questions about what should follow, in turn leading to the design of the T-34.

In April 1937, the commander of the tank force sent to Spain attended a conference with tank works directors, including the boss of the Kharkov Locomotive Factory, also known as Factory No. 183. Mikhail Koshkin was among those attending, as the head of a section within that plant's design department, and eventually he came to be known as the father of the T-34.

Koshkin, who was 39 years old at the time, had what was to the Bolsheviks a pleasing 'peasant' background, had been a party member since 1919 and had a suitable technical education. 'He is a self-starter, energetic, and stubborn,' Koshkin's vetting report noted prior to his appointment as Factory No. 183's chief designer in 1939. 'He is a good organizer and leader, he earned the respect of factory management.'

Discussions with those fighting in Spain told the designers that the Soviet Union's next generation of tanks would have to be more heavily armed (with a 76mm gun in lieu of a 45mm), better suited to fighting in towns, and significantly better protected, able for example to withstand 45mm armour-piercing shots at 1,000 metres.

Work began on two new vehicles. One, the BT-20 (later A-20), would replace the fast tanks, featuring a similar Christie-type 'convertible drive' system; the other, the A-32, was meant to replace the T-26. There followed two years of design bureau rivalry, testing and negotiating between competing military-industrial interests. Many designers thus had a hand in what followed.

The A-20 emerging from the Kharkov plant was a strikingly modern shape. Its hull armour tapered at front, back and sides. The turret also presented few flat faces towards an enemy. Sloping armour, as we've seen, increased the amount that an enemy projectile would have to penetrate, and made it more likely that armour-piercing shot would just glance off.

Its design was all the cleverer because the hull expanded over the top of the tracks before tapering inwards. A British vehicle of the same period, the A13 cruiser tank, which also used a Christie-type suspension, didn't do this, a decision which might seem trivial but which would reduce the hull's width, therefore the size of the turret ring or mounting, which would in turn place greater limits on the size of gun the British vehicle could mount. Indeed, the A-20 and resulting T-34 were significantly broader than the British A13 or Matilda II: 3 metres compared to 2.54m or 2.59m respectively.

There were various reasons for this, but just as German designers were limited by the weight their bridges could take, so the British had a War Office requirement to make tanks deployable on the national railway system. The flatcars were made to fit through Britain's relatively small tunnels while riding on standard-gauge rails. The Russians, with their broad-gauge rail system, had more leeway.

The consequences were felt not just in terms of the ability to mount a larger turret or exploit a wider railway flatcar. When T-34 crewman Sergeant Semyon Aria got a chance to drive a British tank, he was not impressed, noting: 'they . . . would easily tip over on the slopes of hills.' The other thing that he and many other crewmen commented on about Western tanks was that 'they had petrol engines and burnt like torches.'

During 1939 the two rival prototypes evolved. Their development was conducted in great secrecy, and even though the Germans cooperated on some secret tank work in the 1930s, and entered into a non-aggression pact with the USSR in 1939, they knew nothing about the trials of these new machines. Pretty quickly the decision was made to drop the convertible drive on the A-20. That saved 750kg, reduced complexity and gave more interior space for the crew. But as ballistic trials went on, concerns about the power of foreign anti-tank guns led to requests for more armour. The weight reduction gained by one decision would be eaten up by another.

Once again, the eternal designer's trade-off between firepower, mobility and protection asserted itself. A decision to add 5mm to the armour of the A-20 added nearly 750kg to its weight. One to up-armour the A-32 by 10mm increased it by more than 1,600kg. In general, both vehicles got heavier as trials progressed, the A-20 breaking out of its original specified 16.5-tonne weight limit by more than 5 tonnes. The circle was squared in this case by the adoption of a powerful new diesel engine, the V-2. Initially developed for the *Voroshilovyets*, a heavy tractor with industrial, agricultural and military uses, this remarkable powerplant could reliably produce more than 400hp, later being uprated to 500hp.

By a decree of the 19th of December 1939, the parameters for the new vehicle that became the T-34 had been set. The competition between prototypes and designers was over. The A-20 had kept growing in weight, so why not go with the heavier A-32, powered by the V-2 and also with Christie-style suspension? It would be just as agile across country, but protected with sloped, 45mm-thick armour, and pack the punch of a 76.2mm gun.

The same official diktat that set the project in motion also mandated some further improvements to production models. The committee 'consider it necessary to increase space inside the turret to ensure more comfortable placement of the commander and loader'. In response the turret casting was made a little larger, and the radio (for the minority of early T-34s that actually had them) relocated to the hull. There a hull machine gunner, who could barely see out, given the narrow aperture for aiming his weapon, was to make himself more useful as the set's operator.

All of this was achieved covertly, hence the Germans' shock when they encountered the T-34 in June 1941. And there was every reason for their consternation – the breakthrough decision to adopt the V-2 engine had allowed Koshkin's team to resolve the firepower–mobility–protection dilemma. They would be building a tank with a 29-tonne fighting weight, with the firepower to knock out all German tanks then in service, and enough armour to withstand the enemy's main projectiles except at very close range.

There is no doubt that the achievement of these advances was the source of considerable pride to army, party, and the engineering hierarchy. Vadim Malyshev, a senior official and favourite of Stalin's who was used as a troubleshooter on major projects, wrote of the T-34: 'These are not tractors. We cannot crush them with numbers. We need to surpass the current level of vehicles, make the enemy's tanks impotent on the battlefield . . . quality is the path to supremacy.' Malyshev was really highlighting the sense of achievement with the new design. This was mirrored by those lower down the food chain, one crewman opining decades later: 'To tell the truth, we were afraid of being posted to fight in foreign-made tanks . . . Our tanks were the best. The T-34 was a superb tank.'

The Kharkov No. 183 Factory had beaten the others to get the production order, with highly ambitious production targets. In

March 1940, Koshkin and his people presented their prototypes to Stalin in person. Given that their leader was taking the closest personal interest in the project and that war had already broken out between Germany and the Western Allies, the stakes – and indeed the jeopardy – could hardly have been greater.

Orders for the T-34 placed at the end of 1939 gave them just months to deliver what was in Soviet terms a highly sophisticated machine. Factory No. 183 was tasked to build 200 in 1940, rising to an annual production rate of 1,600 by the beginning of 1941. In order to bring the new vehicle rapidly into service, plants in Leningrad and Stalingrad also started production, with the aim that they should make 600 and 200 respectively in 1940, giving target deliveries of more than 800 by the end of that year.

In Stalin's Russia, those running heavy industry held a privileged position. The pouring of molten steel, surging of electrical power, and smashing of production 'norms' set by state planners featured heavily in Communist propaganda. The project of creating world-class tanks could not have been more attractive to a leadership bent on demonstrating the superiority of scientific socialism.

Mikhail Koshkin, and other engineers like him, were, if not celebrities, certainly a sort of higher clergy in this faith of progress through proletarian toil in factories. Design bureaus in the 1930s were hothouses, with the most successful innovators becoming household names: MiG aircraft, for example, being the creation of Messrs Mikoyan and Gurevich. Added to this notoriety were many rewards, from state prizes to access to exclusive clinics or simply getting better food in the works canteen.

There was, though, in the USSR of the 1930s, a fierce culture of repression accompanied by waves of purges and millions being dispatched to forced labour camps or simply shot. Factories, in common with army units, had their special sections of *Chekists* (the word coming from the Russian acronym for the state security apparatus) who were there to root out foreign spies, 'wreckers' and saboteurs. These last terms, ill-defined ways of scapegoating people when crazy production targets were not met, allowed score-settling among rival bosses to turn lethal.

In terms of the Kharkov design bureau, Koshkin need have looked no further than his predecessor to see what could happen. Afanasiy Firsov was purged as chief designer, on trumped-up charges that he had sabotaged the gearbox design of the BT tank series. In Stalin's time, such disgraced individuals were carted off by the secret police and, having once been at the centre of a professional network, suddenly became non-people whose names few dared to utter.

At least Koshkin had the ideal pedigree for a Communist factory boss, being a long-standing party member with no foreign contacts (according to his vetting report) and a perfect exemplar of the 'new Soviet man' in his evolution from peasant origins to heavy industry. Raisa Raikel, wife of one of the designers under Koshkin, gave a vivid pen portrait of him after he arrived at Kharkov, having left his wife and three daughters behind in Leningrad in order to focus on the job in hand: 'Koshkin was ardent. Furious. He lived and burned. This rage helped him develop the tank. He knew how to lead . . . There was nothing worldly about Koshkin. He never thought about everyday life. He was in a hurry. He wouldn't let the cigarette out of his mouth. He sent all the money to his family. Pale. Always hungry. Tired, but there was a fire inside.' This picture of relentless, chain-smoking activity helps explain why the T-34 progressed as quickly as it did. But there were limits to what he could do.

Much of the constellation of plants needed to make the tank lay in places familiar to us from the recent war in Ukraine. Factory No. 183 in Kharkov, Kharkiv to Ukrainians, remains a tank facility to this day, having been an early target for Russian strikes in 2022. The coal and steel needed to make the T-34 came from the Don Basin or Donbas; it was rolled and forged into the tank's hull at a plant in Mariupol. The special glass needed to make its vision blocks was made in Izyum. Perhaps in the mind of Vladimir Putin there is a conflation between the mythic symbols of the war, like the T-34 which sits on memorial plinths across the former Soviet Union, the economic power of the Donbas, and a desire to possess it.

As war approached in 1940, Koshkin was literally consumed by his project. Ignoring the advice of doctors to cut back on smoking, he kept driving himself. When in March a pair of A-34 prototypes left the factory in order to motor to the army's proving grounds near Moscow, he went with them, riding on one of the vehicles. The journey, in which the tanks literally rumbled the 500 miles to the capital, rather than being sent by rail or road transporters, proved to be a triumphant demonstration of their reliability. They ploughed through snowdrifts, roared up the poor roads, and were shown to Stalin and other top officials on the 31st of March.

Koshkin, though, had fallen into an icy river early in the journey. He got pneumonia, and his health went into decline. Despite surgery to remove one lung, he died in September 1940. Whether or not it was an underlying pulmonary condition that claimed his life, the idea that he paid the ultimate price for the T-34 project became part of the legend.

He was succeeded by another senior engineer at the plant, Alexander Morozov, who reaped much credit for the tank's success and ultimately became the Soviet Union's most celebrated tank designer. While Morozov received more recognition during Soviet times, a 2018 Russian film, *Tanks*, lionized Koshkin and was seen by cultural critics as part of Putin's campaign to exploit the war for his present-day political ends.

Major production orders followed the trials in March 1940, and with the design still being tweaked, the plans unveiled at the end of 1939 could not be fulfilled. Rather than making 800-plus T-34s in 1940, 115 were built. Production did start to pick up significantly the following year, but in June 1941 the German invasion naturally enough created its own disruptions, most importantly the overrunning of Factory No. 183 when Kharkov fell four months later. The Germans found sufficient parts there to run off a few dozen T-34s for themselves.

Across the western USSR key industries were evacuated, with dies, tools and expertise moved eastwards. This involved two of the T-34's key final assembly sites (Kharkov and Leningrad) as well as numerous component makers. The T-34 would be built in seven different Russian plants during the war. Some vehicles – made in Stalingrad and Leningrad – would be driven just a few miles from factory to frontline as the sieges there began. In terms of volume, the most significant shifts occurred from Kharkov to Nizhny Tagil in the Ural mountains, establishing what would in Cold War times become the world's biggest tank-production plant, and to Chelyabinsk, similarly far east, where the *Tankograd* or 'Tank City' plant was established.

By 1944, Tankograd would employ 60,000 workers, doing 12-hour shifts seven days a week. The importance of this site was bolstered by the transfer of V-2 engine production there from its previous site in Kharkov. The Zhdanov plant, earlier site of Renault licensed production in Gorky, also grew in importance during the war as its lines were switched from light tanks to making T-34s.

Achieving this eastward displacement of the entire tank supply chain while fighting across the length of the front was in itself a stunning achievement. As hundreds of trains carried machine tools and components off to the Urals, production was bound to suffer. Yet despite this, Soviet industry was able to manufacture 2,939 T-34s in 1942–43 (as well as 3,000 KV-1 heavy tanks and more than 8,000 T-70 lights). The first T-34 came off the production line in Nizhny Tagil on the 25th of December 1941, just a couple of months after the Kharkov staff evacuated there.

The assessment made early in 1944 by the Military Science College in Britain seems only just: 'When it is considered how recently Russia has become industrialized and how great a proportion of the industrialized regions have been overrun by the enemy . . . the design and production of such useful tanks in such great numbers stands out as an engineering achievement of the first magnitude.' It must also be said that the USSR's Allied partners played a key role in sustaining the Red Army's tank force while this eastward production shift was going on.

Under the Lend-Lease programme, thousands of tanks made in the UK, Canada and the USA were shipped across to northern Russian ports. Arctic convoys carrying these precious cargos had to run the gauntlet of U-boats, surface raiders and Luftwaffe strikes flying from airfields in occupied Norway. The first British-supplied machines were thrown into battle in November 1941. Recalling that the Red Army's losses reduced the tank fleet facing the Axis invaders from about 10,000 at the start of Barbarossa to around one-third of that at the end of 1941, the significance of Lend-Lease becomes clear.

During the first half of 1942, when the Red Army attempted its first large-scale counter-offensive (trying to retake Kharkov), British-built Matildas, Valentines and Churchills were a critical part of the effort. The UK and Canada would ship more than 5,200 of them to the USSR, and later supplies by the US, mainly of the M4A3 Sherman, topped 7,000. For context, that combined total of 12,200 Lend-Lease tanks compared to total USSR production of nearly 100,000 during the war.

There can be many debates about Allied aid, not least that the UK supply of 7,000 combat aircraft, or America's gift of 400,000 trucks, ultimately proved of greater strategic significance than the tank shipments. But the most important thing about sending those machines was that they sustained the Red Army at a critical time, after its catastrophic losses of 1941 and before major production from plants in the Urals was underway. Deliveries of the original versions of the T-34, armed with a 76.2mm gun, would peak in 1943, with more than 15,700 made that year. That gives an idea how the losses suffered at Kursk were so quickly replaced.

By then the design was already showing its age, and its limitations were all too evident to the men and women who fought and died in them. For if the story of the T-34 is a remarkable saga of engineering savvy and resilience in production, the entire effort would have been worthless without the user. And in that respect also, the Red Army's *tankists* or crews were quite unique.

One of the great differences between Soviet and Western armoured forces during the war years was the close association between the makers and end users. The 19th Tank Training Regiment, for example, was based beside the Nizhny Tagil plant. On being sent there in 1942, Sergeant Aria recalled that during their 'two or three months of training they also toiled on the assembly line, making tanks'.

Each of the major tank-building factories had its associated training regiment. Posted to Chelyabinsk, another of the Ural sites, Lieutenant Bryukhov explained how it worked: 'As the factory was short of workers I went to work there together with a friend. I quickly learned to operate a semi-automatic lathe and worked for two weeks boring cylinder blocks. We worked for free; in fact, we only received a lunch card.'

Thus, it was usual for trainee crews to assemble the vehicles they would eventually take into battle. Drivers-to-be might find themselves building gearboxes, or operators installing their radio sets. It was a neat way to help the plants through labour shortages following the move east, while giving tankists, some of whom might have come from the most backward rural communities, an intimate knowledge of the machinery that their lives would depend upon.

The courses generally lasted months and as they came to an end a group of trainees would be assigned to a bloc of vehicles, and on the appropriate day a new tank battalion left the factory gates complete. Georgi Krivov would later recall: 'When my crew was given our first tank, it was actually nothing but an armoured box. We watched the assembly process for some time and then went to have dinner. In an hour we were back, but our tank was missing . . . the assembly process had already gone as far as mounting the turret.' In some cases, the novice tankists helped build the *actual* vehicle they would take into battle.

Veteran testimony and official reports from the 1940s alike testify to what a difficult experience operating the new T-34s could be. Koshkin's 1940 rush job may have been a strategic necessity but it produced a vehicle with some glaring faults, despite the general soundness of its design.

Drivers struggled to change gear with the transmission, one noting that 'it was impossible to shift the gear lever with one hand, and I had to help myself with my knee.' Some resorted to striking that selector with a mallet, others got the help of the radio operator sitting next to them. There was more than one way to assist the driver, one crewman noting: 'since his hands were busy I'd roll cigarettes for him, light them, then put them in his mouth.'

The original four-speed transmission attached to the V-2 engines in the T-34 and KV-1 was the bane of many crews' lives. The testing centre at Kubinka was doing no more than stating facts when it reported in 1942 that the gearboxes 'do not fully meet the requirements of modern combat vehicles, being inferior to the tanks of the Allies as well as the enemy's tanks'. Later that year a new five-speed box went into production.

Another area where the T-34 drew unfavourable comparisons was its noise. It wasn't just the diesel engine, but the tracks and running gear more generally that caused it to make such a racket. While the tank's wide tracks were praised for offering low ground pressure, the designers' use of a plate on the hull's side to keep the track pins from slipping out (rather than the clips used on Western tanks) created a lot of noise. The exigencies of wartime production, when for example the Stalingrad factory produced road wheels without rubber on their rims because the siege interrupted the supply, just added to the cacophony. Listen to a restored Panzer III or IV and it is remarkable how quiet their Maybach engines are. T-34 crews felt at a definite disadvantage when stalking them. The Lend-Lease British Valentine earned Soviet plaudits for its quietness also.

If the driver's labours proved physically taxing, those of the commander could addle the brain. Lieutenant Bryukhov, quoted earlier describing Prokhorovka, felt that 'if you were the commander of a T-34/76 you had to do everything yourself – fire the main gun, lead the platoon over the radio, everything.' Quite a few of the issues were fundamental to the original design.

The turret fitted to the first few thousand T-34s was undoubtedly the worst aspect of that. It was weak conceptually, in giving all the work to two crew, and it was doubly so in its realization. Soviet armour had a lower zinc content than that used by Lend-Lease tanks, which gave rise to more spalling or flaking off of pieces of metal on the inside when it was struck. The casting technique used for the T-34's turret (but not the hull, which used rolled and welded steel) exacerbated this issue. This factor, along with the tactical use of vehicles (where the upper part was more likely to be exposed to an enemy), led to a conviction among T-34 crews that they were more likely to die in the turret.

As for fighting with the vehicle, the commander had a single viewing periscope to find targets with. This was quite different to the boss of a Panzer IV, who was surrounded with an array of periscopes offering 360-degree vision. In early-model T-34s it didn't even use glass prisms but polished steel to reflect the battlefield panorama, with dirt or fogging often obscuring the view.

Resorting to the simple expedient of putting your head out to have a look around was also hard, because commander and loader shared a large, slab-like double-width hatch that, one user recalled, 'was heavy and hard to open. If it got jammed, that was it, no one could bail out.' It was also drilled into them that the hatches had to be closed down in combat.

So, when rolling into battle, a commander had very poor situational awareness, bounced around by the terrain while trying to keep his eye steady on something through his periscope or aim through the soda straw perspective of his gunsight. An understanding of this is key to explaining the dramatic disparity in T-34 losses when facing the Germans, for example at Prokhorovka.

Lieutenant Georgi Krivov was sent as a replacement to one of the 5th Guards Tank Army's brigades after that Kursk slaughter.

Going into action for the first time, he summed up well the dilemmas of a commander when sent forward but barely able to see: 'Suddenly several anti-tank guns fired at us simultaneously. I was trying to adjust the sights to target one of them, but in vain. It was impossible to aim at anything on the move, we needed a stop. But I remembered the words of my friends: once you stop, your time is up!'

Tank commander's black eye, it should be noted, is a perennial issue. The neophyte soon learns to keep their face a little way back from the sight, particularly at the moment when the main gun fires. But the issues described by Krivov were particularly bad, related as they were to the more basic question of being able to see who was shooting at you before they hit.

Like the gearbox issue, changes were made to the turret design, with a larger, hexagonal shape as well as individual hatches for commander and loader. Nicknamed the 'screw nut' by crews, this design was soon adopted in the Ural factories. However, the problems posed by two-man turret crewing and the lack of quality vision devices remained.

When an officer was given added responsibilities, commanding larger groups for example, the limitations of the T-34's turret were such that he often decided to go elsewhere. Arkadi Maryevski recalled: 'Later on, after getting promoted to a T-34 company commander, I never sat in the turret and usually drove the tank.'

While drivers were also supposed to go into battle with their hatch closed, this regulation was widely ignored. And in this case, leaving ajar the armoured plate directly in front of the driving position did afford a panoramic view of the battlefield, albeit without the added height gained from being up top. The idea of a company commander (usually with eight to 12 vehicles under them) leading from this position is a remarkable one. It is fair to say that it would have been unheard of in the German or British armies. It serves as an indictment of the T-34's design but also as an acknowledgement of certain realities.

In the Red Army context, command had to be exercised quite

differently – so the idea of a leader literally putting himself in the driving seat then saying 'follow me' makes more sense. Only a few of their tanks were equipped with radios when Operation Barbarossa began. An official form tells us the exact figure: 221 out of 671 T-34s inspected in June 1941 had them. Of course, that paperwork does not say whether those sets were actually working.

Soviet doctrine held that 'linear' tanks (as those without radios were called) had to be directed by flag signals, 20 of which were taught to trainee tankists. But as we have seen, it also mandated that crews had to be closed down in battle, which meant no waving around of coloured flags.

Over time, efforts were made to equip all Soviet tanks with radios. Britain performed a vital role in this, sending untold thousands of valves, transistors and other components via the Arctic convoys. The increase in effectiveness made possible by those Lend-Lease supplies is very hard to quantify, but Allied support in this area undoubtedly helped in the general improvement in the Red Army's tank force during 1944–45.

It might seem surprising, given all of the limitations of early-model T-34s, that Red Army tankists did not express a strong preference for Allied tanks. Certainly, they were very much aware of the creature comforts, one recalling: 'I had a look at an American M4A2 Sherman. My God! It was like a hotel inside! It was all lined with leather so that you didn't smash your head.'

However, Soviet crews knew the limitations of designs like the Matilda or Churchill, considering them under-gunned and too slow. They were also convinced that they stood a better chance of survival in a diesel-powered tank than in those petrol-driven models. While diesel fuel has a higher flash point, and its fumes are far less combustible, this was not quite the advantage they perceived, since once alight the T-34's fuel tanks also burned very well.

However, the Valentine remained a favourite because it was small, quiet, well armoured for what was a 'light' tank in Soviet terms, and, when armed with a 6-pounder (57mm) high-velocity gun, packed quite a punch. The Sherman too had its fans. Newsreel footage shows that those two types went all the way to Berlin, whereas others faded away.

Despite the limitations of the T-34's design – and these largely related to the turret design – it remained a very well protected vehicle with good firepower, and it had real advantages, particularly during the 1941–42 period when German crews found it so hard to counter. Among the countless dubious claims of Tigers and other panzers destroyed, Red Army crews did of course manage to use T-34s successfully on many occasions, not least in the late summer of 1942 to cut the Germans off in Stalingrad.

While Soviet propaganda eschewed the more florid Nazi depiction of tank crews as modern knights, a steady stream of awards and news stories inspired 'socialist emulation' of their feats. The citation for their highest decoration, the Hero of the Soviet Union gold star, awarded to Major Nikolai Bezrukov, gives a sense of it: 'On September 7th, 1943, during the battles for Buryn, he corrected the fire of battalion tanks from atop his tank. He was heavily wounded while doing this, but did not leave the battlefield until the enemy was defeated. In these battles, comrade Bezrukov's battalion destroyed up to 10 enemy tanks and SPGs, significant amounts of other armament and vehicles, and 300 fascists.' It is worth noting a few things: the kills claimed of enemy armour are not extravagant; the wider toll of 'fascists' killed is deemed relevant; and Bezrukov evidently felt unable to lead his battalion from inside the tank, hence directed its fire from on top!

Sifting accounts of the Red Army's war, it's clear that while many tankists considered German armour their most dangerous enemy, and therefore a priority target, there was an understanding that they could not expect to beat them tank-to-tank, unless luck was on their side. The standard ammunition loadout for a T-34, 75 high-explosive rounds and 25 armour-piercing, underlines the main role of supporting infantry. German ideas of striving for vastly disproportionate tank kill ratios appealed ideologically to the Nazis but were also necessary because the Bolsheviks could outproduce them by such a big margin.

A Soviet examination of armoured doctrine, produced by the People's Commissariat of Defence late in 1942, makes explicit some of the differences, and underlines the degree to which they expected their tank crews to be harmonious players in the orchestra of war rather than dazzling soloists. It castigates armoured commanders who got 'diverted into combat against enemy tanks and artillery', noting that their primary purpose was 'destroying enemy infantry'.

This critique of tank troops' leadership argued that they failed to use radios properly, assuming they had them, meaning they lost control once their vehicles were committed to action. In the committee's view, they 'have no influence over the course of the battle . . . the units, lacking any direction, become lost and confused on the battlefield.' These problems were not caused by their tank designs, but clearly they exacerbated them.

You might think that the committee's observations were ignored, because these faults still appear very evident in the tank units' lemming-like drive to the anti-tank ditch on the 12th of July 1943. However, it is fair to say that the Red Army's coordination of its different battlefield arms had improved markedly by Kursk, where the integration of field engineers, artillery barrages, anti-tank brigades and armour made their defences so formidable.

Certainly, the Germans had realized that warfare was changing, due in part to the way the Red Army was fighting, calling into question whether the panzer force could still be a decisive arm on the battlefield. 'The tactics that led to the great successes in 1939, 1940, and the first part of Barbarossa in 1941 must now be considered as outdated,' wrote Lieutenant General Fridolin von Senger und Etterlin in April 1943, while commanding the 17th Panzer Division. Breaking into Russian defensive belts was proving unsustainably costly, he believed: 'the tank is all too often vulnerable when the enemy deploys a more powerful anti-tank weapon that can defeat our armour.'

It's important, in surveying armoured warfare over more than a century, to understand the significance of this time. Germany's enemies had by 1943 learned how to counter blitzkrieg tactics more effectively. They had invested in anti-tank weapons, and learned many a bitter lesson in the need for teamwork, particularly between infantry, armour and artillery.

All of the main belligerents – including Germany, Britain and the US – reorganized their armoured divisions during 1942–43, reducing the number of tanks, bolstering other arms. The idea of a great armoured steamroller crushing everything before it, gaining great advances, ebbed.

Much as Soviet weaponry was developing apace during 1943, with powerful 85mm and 122mm anti-tank guns coming on-line, the T-34 with its 76.2mm gun would remain the mainstay for a while longer. And having considered the degree to which those who operated them were trained and often commanded quite differently from the armoured forces of other countries, it is only fair to consider one other aspect which was quite different: the degree of coercion they faced when going into battle.

There are stories of officers on the Allied side drawing their automatics and threatening people – albeit a very rare occurrence. In the Wehrmacht, those like the officer threatening a battery commander in Normandy with the chance of a medal or death may not have been quite so rare, but this did not reach the level of menace in the Red Army. And of course the Germans readily consigned 'subject' peoples to slave labour facilities and Jews to concentration camps. In Stalin's Russia, though, there was a whole paraphernalia of terror directed at their own army, from the Chekists in each unit to 'blocking detachments' ordered to shoot retreaters, penal battalions sent on suicide missions and the whole ghastly system of forced labour.

Lieutenant Bryukhov, already cited on Kursk, had the temerity to question an order that he felt would result in the certain death of his platoon, drawing the response from his divisional commander, 'Shut up or I'll execute you! Fulfil the order!' Where things did go badly wrong, the search for culprits, wreckers, mirrored the purges of the 1930s.

Another young officer in the 5th Guards Tank Army, returning on foot after his T-34 had been knocked out, found himself being cross-examined by a goon from 'counter-intelligence'. 'Did the tank burn or not?' the shaken commander was asked. 'If it didn't, then you have to be court-martialled for abandoning your vehicle.' Sometimes the grounds for drastic action seemed grotesque. One crewman recalled long after the war: 'Three drivers were executed in our unit between Vitebsk and Polotsk. They exposed their side armour to the Germans.' The supposed crime was turning your weaker flank towards the enemy so as to make it more likely you'd be knocked out.

The huge losses in tanks, particularly during the early months of Germany's invasion, could only be explained in the eyes of the party leadership and *Smersh* counter-intelligence units by malingering, cowardice or treachery. And the wilful abandonment of a tank, as with other state property in the Soviet Union, was regarded as a particularly heinous crime.

Naturally this paranoia emanated from those at the top and worked its way down. In August 1943, following reports that 326 out of 400 tanks in the Stalingrad front were out of action, 200 because of breakdowns, Stalin issued the following order: 'I consider such an impermissibly high percentage of tanks lost due to technical defects unbelievable. The *Stavka* [supreme high command] is seeing here the presence of hidden sabotage and wrecking on the part of a certain number of tankists, who are trying to avoid battle and abandoning the tank on the battlefield by searching for isolated minor bugs in the tanks or deliberately creating them.'

Where suspicion fell on those higher up the chain of command for a failed operation, survival often depended on throwing the blame onto others. Lieutenant General Rotmistrov, under suspicion following the Prokhorovka debacle, escaped as the tide of battle following the Kursk offensive began to flow in the Red Army's direction. But he made sure to pen a memo to Marshal Georgi Zhukhov in which he blasted the 'conservatism and conceit' of the USSR's tank designers, reporting: 'our tanks today have lost their superiority over enemy tanks in armour and armament . . . The German tanks' powerful weapon, strong armour and excellent optical equipment place our tankists in a plainly unfavourable position.'

Relief, though, was on its way. Production was started of the IS-2, a heavy tank with a mighty 122mm gun able to duke it out with the Tigers. Meanwhile, the designers at the Zhdanov plant in Gorky were preparing a radical upgrade of the T-34. By fitting the potent 85mm anti-tank gun in an entirely new turret they solved several problems at once. Not only would the T-34/85, as it became known, have a significant upgrade in firepower, but the issues facing its commander would also be addressed. With a turret crew of three, a gunner and loader could take care of engaging the enemy, while providing vision blocks built into a cupola gave the man in charge a significantly better view of the battlefield.

This was a far bigger change than the fitting of a long 75mm gun to the Panzer IV – it was a reimagining of the design into something that some armies would continue to use 50 and more years after the Second World War. I was unsurprised to see a T-34/85 in action in Bosnia in 1992. It was a machine that had achieved worldwide export success, notably during the early days of the Cold War.

As a weapon against the Nazis, it was in early 1944 that the first T-34/85 went into service, and really only in the final months of the war that sufficient thousands had been delivered to make it the fleet's mainstay. It was the original design, subsequently dubbed the T-34/76, that proved to be the archetypal Red Army tank of the Second World War.

Just over 35,000 of that earlier variant and 23,000 with the 85mm gun were delivered prior to VE Day. Out of this combined total of 58,000 it has been estimated by historian Steven Zaloga that 44,900 were knocked out. He has argued that the T-34 deserves the dubious accolade of 'most knocked out' tank in military history. Russian historian Valeriy Zamulin has calculated that the Red Army suffered a tank loss rate of 427 per cent, i.e. that the armoured forces facing Germany on the 22nd of June 1941 were lost more than four times over. These statistics certainly bring home the brutality of the Eastern Front tank war, and its very unequal nature.

While both sides exaggerated kill claims, they help us understand that loss rates of 15:1 or even 20:1 in the invader's favour did occur on quite a widespread basis. Given the Communist obsession, fetish even, with molten metal, production and heroic sacrifice, it might be argued that a character like Stalin would regard the technological superiority of German tanks as an irrelevance. The T-34, produced in such large numbers under adverse circumstances, crewed by people with so little hope of survival, represents a perfect symbol of Soviet resilience and resistance.

Of course, the Soviet Communists would have been less happy conceding the role played by Allied Lend-Lease shipments, keeping the Soviet Union fighting when the Nazis were at the gates of Moscow, then motoring their supplies all the way to Berlin.

If Soviet production was impressive, especially given the disruption of 1941, that of the United States was to prove even more so. It would not only play its part in equipping the Red Army, but in furnishing the Allied armies with the materiel needed to sap the Germans with multi-front warfare. This triumph of capitalism was embodied by the Sherman tank.

Sherman M4

Entered service: 1942
Number produced: 49,000 (of various marks)
Weight: 30.2 tonnes (M4A1)
Crew: 5
Main armament: 75mm gun
Cost: $35,000 (the original spec, equivalent to £8,750 in 1942, which in turn is around £345,000 in 2024. However, the US government later assessed the delivered cost including some equipment they added to be $47,800 per vehicle)

Sherman

Night operations were never simple, especially in a featureless desert. The tanks of the 9th Armoured Brigade had been ordered to move up in darkness, their drivers straining their eyes to stay within cleared lanes marked by white tape. The business of that evening was breaching a defensive belt, starting with a deep minefield.

The cavalrymen were supporting the 2nd New Zealand Division during the early hours of the 2nd of November 1942. A crushing artillery barrage had been followed by mine clearing – carried out by men probing the sand with bayonets, and by a new invention, the Scorpion. It was an early example of a 'Funny', a tank adapted to a specific purpose, in this case a flail or mechanical drum that rotated to its front, thumping the sand with dozens of heavy chains in order to set off mines.

With the initial lanes cleared, 9th Armoured was to advance about an hour before daybreak. It was a motley formation, shaped by the exigencies of war. The senior regiment, 3rd Hussars, were an old regular mob, experienced soldiers. Its other two, however, were reserve cavalry, yeomanry from Warwickshire and Wiltshire, whose only experience of desert tank soldiering had been an aborted attempt to advance in this same axis on the 24th of October. The hope was that the Hussars would steady the others.

This brigade – which suitably enough, being the 4th Cavalry Brigade pre-mechanization, had a white horse as its symbol – was mixed from the point of view of equipment too. Its 90 tanks were made up of four different models, using three kinds of fuel, and mounting nine types of gun or machine gun. Such were the fortunes of war that this logistician's nightmare had been tasked with a key role at this moment, the climax of the Battle of El Alamein.

It had also been given a most coveted bit of kit, for this was to prove the Sherman tank's baptism of fire.

Lieutenant General Bernard Montgomery, leading the 8th Army at Alamein on the coast of Egypt, had told the armoured brigade's commander, 'If necessary, I'm prepared to accept 100 per cent casualties in both personnel and tanks.' Infantry alone could not breach the German defences on the ground that night, rising gently towards the Tel al Aqqaqir, a ridge barely distinguishable from the surrounding desert. The enemy were defending their minefields and wire with machine guns and 50mm Pak 38 anti-tank guns; further back, if the British made it that far, several of the dreaded 88s and indeed armour awaited.

There were 36 Shermans in the brigade – the regiments having a squadron each of light tanks, Grants (American-made mediums with the same 75mm gun as the Sherman) and the new vehicle. The Kiwi infantry was glad for the backing, one recording: 'A long line of tanks breaks through the gloom. One, two, three – no use trying to count them as they appear and vanish in the gathering dust.' The cavalry was going to work.

Initially, plunging into the darkness, they surprised Germans in their forward positions. As men leapt out of their trenches, the chatter of machine gun fire and whiplash of tracer rent the darkness. It didn't take long before the forward vehicles came across some of the German anti-tank guns, crushing the Pak 38s under their tracks, scattering the crews. But after an hour or so of the initial advance, everything began to change. Some of the tanks, taking a shortcut, drove onto mines, knocking out four. And as the first glimmer of dawn broke, the crews, who were attacking from east to west, realized that the rising sun had silhouetted them against the skyline. 'Nothing happened for a few minutes, and then hell was let loose,' Captain Clive Stoddart of the Warwickshire Yeomanry wrote home. The 88s had begun to bark.

Some of the British tank commanders referred to the 'white tennis ball', the 88mm projectile roaring past them. Others recalled the green tracer in the shell's base, and for those who had them pass very close, the rush of air and shock of concussion as it went by. Too many, though, found their targets. '0700 very heavy fighting,' Lieutenant Henrik Karsten of the 3rd Hussars would write in his journal that evening. 'God knows how I got out alive. I was expecting to be blown sky-high *any second. All-day* shelled. Heavy losses. Very few tanks left.'

Stoddart, knocked out of one tank, crawled over to a disabled Sherman, climbed into the turret and, finding the radio working, tried to raise his commanders to ask them what to do. To his consternation, a German voice replied: 'I hope you haven't been in trouble, but isn't it nice to see all your new toys going up in smoke!' The few surviving tanks lobbed shells at the German anti-tank guns on the ridge, trying still to improve their odds. But then some panzers, approaching from dead ground behind that height, opened up on the British.

Monty would describe the 9th Armoured Brigade's action as a 'blood sacrifice'; the commanding officer of the 3rd Hussars called it 'a suicide mission'. It was never going to be pretty, the 8th Army commander had told them that, but he reckoned the losses would be worth it if 9th Armoured could 'kick open the door', and a follow-on echelon, the 2nd Armoured Brigade (also partly equipped with Shermans), could pour through the gap and into the enemy rear.

Ninth Armoured Brigade had gone into action at the start of Alamein with 128 tanks, losing some vehicles in its brief advance of the 24th of October, starting the attack of the 2nd of November with 90 of them. By the end of that day there were just two dozen runners left. Some 270 of the brigade's 400 crewmen were killed, wounded or missing also.

The brigade commander expressed understandable bitterness that the follow-on forces of 2nd Armoured were too slow to exploit such a hard-won advance. They responded that the cavalry had failed to create a gap. These recriminations were within days to give way to a wider argument among Montgomery's querulous commanders about whether the armoured divisions had failed more widely to exploit the victory of Alamein, allowing General Erwin Rommel a chance to extract his surviving forces.

The bigger picture was that the 8th Army had just achieved a signal victory. The attack of the 2nd New Zealand Division, after days of pounding, and supported by the cavalry, had proven to be the decisive moment, the last straw for the Afrika Korps, for on the 3rd of November Rommel had conceded the game and signalled home his intention to retreat.

While the taking of the Tel al Aqqaqir ridge was only achieved on the 4th, the tanks had broken into the position and significantly undermined the German defence. History does not record whether they knocked out any panzers, but they did overrun forward positions and disable quite a few anti-tank guns. This in itself was something of a revelation to Monty because Britain's desert army had for years fielded tanks with 2-pounder guns that fired a solid shot, largely ineffective against German anti-tank gun crews.

Even before Alamein, American-made M3 Grants had arrived, equipped with a 75mm weapon able to fire high-explosive shells. These 6.7kg projectiles blew up, shattering their steel casing into hundreds of pieces of shrapnel. They could shred enemy gun crews thousands of metres away. Even those serving 88s could no longer assume that they could outrange their enemy or that they would survive his wrath.

Grant tanks mounted this powerful armament in their hulls, so the whole tank had to be turned towards the target. But now the Sherman had appeared, US industry having surmounted the problem of making a turret big enough to take the 75mm gun, and it was truly a game changer.

Just days after Alamein, as he pursued the broken Axis forces westwards, an exuberant Montgomery had signalled London that, as far as tanks went, 'the 75mm gun is all we require.' The importance of the moment had not been lost on the enemy either. A few days into the Alamein battle, the staff of the 15th Panzer Division reported to Berlin that the 8th Army's 'new tanks can open fire at ranges between 2,000 and 2,300m without being effectively combated by most of our tanks'. The desert armoured war had been changed.

Elements of three British armoured brigades had fielded Shermans during the Alamein break-in. These vehicles were part of a shipment of more than 300 that had reached Egypt in September 1942. It was a politically symbolic gesture, US President Franklin Roosevelt having decided that, after the fall of the British fortress of Tobruk three months earlier, something had to be done to boost the fortunes of the 8th Army.

Since production of the Sherman was just getting underway, the White House had directed the US Army's only armoured division to give up its brand-new vehicles so they could be sent to the Brits. It directly set back the formation of America's new armoured corps, but such was FDR's commitment to his allies. The journey hadn't been easy either, one of the ships carrying the new tanks being torpedoed in the Caribbean. Sailing around the Cape (rather than through the Mediterranean, which was still considered too dangerous), the convoy had unloaded at one of Egypt's Red Sea ports, with the vehicles taken forward by rail from there.

British crews had just a few weeks' training on the new machines before the Alamein battle started. But they were immediately impressed. Corporal Alfred Court, a driver in 3rd Royal Tank Regiment, said: 'you could get a fair speed out of it . . . the power-to-weight ratio was fantastic.' Its American builders had taken great pains to build the Sherman to high standards, and Corporal Court and others reaped the benefits in terms of suffering far fewer breakdowns than with British tanks: 'we went many, many miles without any help at all, they were marvellous.'

For others it was the turret-mounted 75mm gun that made all the difference. Lieutenant Allan Waterson of the 2nd Lothians and Border Horse saw how the trajectory and range of that weapon opened up all sorts of new possibilities: 'you could get three rounds in the air before the first had landed.' He also felt it was far more comfortable to operate than the British Crusader he had previously crewed. 'There was plenty of room for all your bedding, equipment, loot, a lot of tanks had livestock on the back of the tank, plenty of room for ammunition, and plenty of room for surplus rations.'

For those who transferred from the M3 light tank or Crusader, the key thing about the Sherman was its firepower: two machine guns for short-range work, and the 75mm main gun for longer reach. Sergeant Douglas Covill of the 10th Royal Hussars later recalled, 'now this was the first time that we'd had a gun that was equivalent to the Germans.'

As we have seen, the enemy was quick to feel this change at Alamein. Over time the tactical consequences of this were absorbed by General Rommel, who had charted the deterioration of his battlefield advantage from Alamein onwards. Several months later, after the defeat of a major attack on Montgomery's troops in March 1943, he wrote: 'the British shot up our tanks, machine gun nests and anti-aircraft and anti-tank gun positions at a range at which our own guns were completely incapable of penetrating their heavier tanks.'

Given the capabilities of this new weapon, it might be asked whether Montgomery threw away that first precious shipment of Shermans, sending them up against minefields and 88s on the Tel al Aqqaqir ridge at Alamein. The breakthrough of a defensive belt with its mines, wire, dug-in infantry, artillery and anti-tank weapons is just about the toughest of missions for armoured forces; and that has remained a constant from Cambrai in 1917 to the Orikhiv battle of 2023. The coordination of all arms is critical, and even small errors in timing can wreck the operation, hence the anger of those in the mauled cavalry regiments that the 2nd Armoured Brigade didn't move up forward more aggressively.

But in considering the Sherman's baptism of fire, and indeed the turning point of the Alamein battle, we need to remember too the awful, attritional nature of warfare in the machine age. Montgomery knew the cost would be very high; he had said before the assault that he might have to sacrifice the entire 9th Armoured Brigade. So the key question for him was whether those lost tanks and their crews did enough to break down the German defences in that sector, so that others might then renew the attack and break through.

His verdict was that this was a necessary and important loss rather than a forlorn one. Writing in 1967, Montgomery reflected: 'If the British armour owed any debt to the infantry of 8th Army, the debt was paid on 2nd November by 9th Armoured Brigade in heroism and blood.'

The tank used in this feat was the product of American heavy industry that was, late in 1942, just gearing up for production on a massive scale. As a machine it combined available technology in the interests of getting rapidly into service and ensuring reliability. This meant that like many weapons produced in the Second World War its period of supremacy was relatively short-lived, perhaps 12–18 months, before German developments changed the calculus.

It was no accident that the first people to use the Sherman in battle were British, nor that the diesel version of the tank would prove to be a success in the Red Army. For in 1942 the United States was trying to expand its own forces at breakneck speed and there was an explicit understanding in Roosevelt's White House that, while it did so, the prodigious capacity of American industry should be mobilized in the wider interest of defeating the Axis powers.

Sitting in front of microphones on the 29th of December 1940, President Roosevelt looked around the room, waiting for his cue. It was a radio broadcast, but like many of his addresses to the nation termed 'fireside chats' because they involved talking to citizens about government policies using layman's language, it was also being filmed.

At that stage the United States had not entered the war, and a great many Americans wanted it to stay that way. The bow-tied president began, 'this is a talk about national security', then launched into his argument for a big shift in national priorities. Describing Britain's struggle against an 'unholy alliance' (the Axis, so-called, of Germany, Italy and Japan), Roosevelt argued: 'Our own future security is greatly dependent on the outcome of that fight.' Challenging the stateside isolationists directly, he continued, 'Our ability to "keep out of war" is going to be affected by that outcome.'

Thus his argument, not unlike those heard since 2022 about sending weapons to Ukraine, was that US security was directly linked to events far away. In order to keep the war from reaching American shores, the US would have to step up its production of weapons greatly: 'Democracy's fight against world conquest is being greatly aided, and must be more greatly aided, by the rearmament of the United States and by sending every ounce and every ton of munitions and supplies that we can possibly spare to help the defenders who are in the front lines.'

In a phrase that was to define Roosevelt's approach to the war, he continued, 'We must be the great arsenal of democracy.' Two and a half months later he would sign into law the Lend-Lease Act under which huge supplies of weapons would flow to the UK and its allies, and after Hitler's invasion of the USSR to them as well.

Although Lend-Lease would embrace everything from warships to fighter planes, the Sherman tank would become one of its most important items both numerically and symbolically. Initially, US plants began supplying the M3 medium tank (termed the Grant in British service) and M3 light tank (the Stuart or Honey to its British crews) under this scheme. In railroad plants stretching from Ohio to the Pacific seaboard – such as Lima Locomotive Works, Pacific Car and Foundry, and the Pullman-Standard Car Company – heavy engineering expertise was switched to making tanks.

However, the most important production base was to prove

Detroit, centre of the automotive industry. In the summer of 1940, six months before that presidential fireside chat, it was decided that a new plant would be needed to bring about a dramatic increase in production. A site was selected in a place called Warren Township, 11 miles from the city centre, for what would be called the Detroit Tank Arsenal.

The first tank made there rolled out of the gates in April 1941. There would eventually be five production lines running, peaking in 1944 with thousands made *monthly*. The Detroit Tank Arsenal was Chrysler's baby, and they ran it with all the same efficiency that they had brought to automobile production. In the period of four and a half years from the first production model appearing to late 1945, more than 25,000 tanks were made at that one facility, 18,000 of them Shermans. And, of course, the US had several others running too. By the end of the war, 49,000 Shermans had been built.

The path to these extraordinary production figures, in common with the evolution of the US Army's ideas about how to use its armour on the battlefield, was initially slow, being hampered by bitter turf fights between the key players. On the industrial side, these arguments pitted the War Department against civilian advisers close to the White House who believed the Office for Emergency Management rather than the military should be co-ordinating this huge shift in industrial priorities.

Meanwhile the military also had to achieve a dramatic call-up, from 200,000 troops in the summer of 1940 to 1.4 million one year later. Enormous targets were set (for example, for 120,000 tanks to be made and 8 million people to be in uniform) that were never met, but nevertheless did bring about the most extraordinary increases in output, from weapons factories to the military training facilities.

When it came to tanks, designs evolved through the interaction of the army, the Ordnance Department, industry, and the proving ground at Aberdeen, Maryland, where new ideas were tried out. Because of this it is sometimes said that the Sherman lacked an individual designer. Rather, there were some similarities with the German system, where the military men set the parameters for what the vehicle was meant to do, the Ordnance Department (like the Heereswaffenamt) translated these into practical plans and industry responded.

None of this was helped, as Roosevelt and others watched the world drifting into war, by the US Army's chequered history with tanks. The First World War Tank Corps, formed with Renault lights and Mark V heavies, was disbanded just three years after the Armistice, much to the dismay of two of its officers, George Patton and Dwight Eisenhower. During the inter-war years, while America's automobile and aviation industries grew apace, the army put little effort into armoured vehicle development, the few it did make being manned by the infantry.

If one had to credit one name with the Sherman's paternity it would be Major General Adna Chaffee, who was made boss of the armoured corps when it was re-established in 1940. Having both infantry and cavalry experience, he would play a key role in defining US Army ideas about armoured warfare as well as the machines they would need to fight it prior to his death in 1941.

The M4 Sherman evolved from the M3, which, logically enough, grew out of the M2 programme. As Washington bureaucrats struggled to make the 'arsenal of democracy' concept a reality, evolution and the use of off-the-shelf components played an important role because, just like the designers of Britain's Mark IV in 1916, that would speed things up. Corporate America, with its concerns for mechanical reliability and the bottom line, also favoured this approach: essentially the Ordnance Department told industry it wanted as many tanks as it could make as quickly as possible, at a price of $35,000 each.

In choosing a powerplant, American designers had no counterpart to the powerful Soviet V-2 diesel. An aero engine, the Wright R-975 Whirlwind, had thus been installed in the M3 tank, and would equip the early-model Shermans also. This radial engine (so called because the nine cylinders were arranged in a circle around the shaft that harnessed their energy) had been around since the early 1930s, powering planes like the Electra airliner.

Lodged in the back of a tank, the Sherman version of the R-975 would eventually squeeze a few more horsepower out of it, as they developed it from the 300hp version used on the M3. For desert warriors, the rotary fan used to cool it sucked air through the crew compartment, which chilled the occupants nicely. There were downsides too: it was very thirsty, and the engine's radial design meant that the power shaft, which would be connected to a propeller on an aircraft, was in its centre, or a couple of feet off the floor of the tank. This, and the decision to drive first the Lee and then the Sherman from sprockets at the front, meant that the vehicles had to be significantly taller than many others because of all the machinery that had to run from the engine compartment to the front, beneath the crew.

So the Lee/Grant was 3.12 metres high, and the Sherman 2.75m, compared to the 2.68m of the Panzer IV. Again there were pros and cons: it made for a spacious interior, but also a bigger target. And a larger internal volume, as we know, had significant implications for the weight of the vehicle for any given thickness of armour.

In terms of armour, the Sherman had around 50mm on its hull front, but raking this at 56 degrees made it the equivalent of 90mm to a projectile hitting it. Side armour was less impressive, being 38mm thick and unsloped. The welding of additional plates on the side to protect specific weak points would prove controversial, some crewmen claiming they just gave their enemy a better aiming point.

The suspension, like the engine, relied on an off-the-shelf solution. M2, M3 and early-model M4 tanks all used paired road wheels on tilting bogies, a design taken from an agricultural tractor. It was an available, robust solution, though it made for a rough ride over bumpy terrain.

As for the main armament, that too was a ready-made design, a 3-inch or 75mm anti-aircraft gun adapted for vehicle use. Its performance came as a revelation to British Grant crews in the desert battles of 1942, and with a few modifications it was adapted for the Sherman. The big difference between the M3 and M4 designs was that, by the time the Sherman went into production, US firms had the ability to mill the bigger rings needed to mount a larger turret. As the British knew, and the Russians also when going from the T-34 to the 85mm gun, a larger mounting ring opened the way to a bigger, heavier turret, and with that to packing a more powerful punch.

In time all the main elements on the tanks at Alamein – engine, gun and suspension – would change as the Sherman evolved. Under the US system for naming these versions, a bewildering variety of letters and digits were added to denote these evolutions.

As major production of the Sherman picked up at the various plants, the shortage of suitable motors meant a variety of engines were tried. A version (the M4A2) with twin Detroit Diesels slaved together proved very popular with crews, not least because this fuel was less likely to catch fire than the gasoline used by the R-975. The diesel version also delivered the best economy, 1.1mpg, about 20 per cent better than the gas-burning predecessor.

While the diesel appealed to tank soldiers everywhere, there were serious logistical implications. In Germany and Britain there were often shortages of suitable fuel. The Red Army had no such problem and thus took Lend-Lease M4A2s with alacrity. The US Marines got this version because the army didn't want it, though it's also been claimed that there were logistical advantages to using the same Detroit Diesels that powered thousands of their landing craft.

During the summer of 1942 other models went into production also, one with twin Ford V8s (the M4A3, forming the mainstay of the US Army's tank force) and the one that would end up being the main British and Commonwealth model, the M4A4. The latter's plant consisted of five petrol engines joined together to make a 30-cylinder monstrosity – the hull had to be lengthened to fit it in.

When Britain's liaison officer in the US, Brigadier G. MacLeod

Ross, was told in November 1942 that the entire output of the Detroit Tank Arsenal would be switched to this model and all allotted to the UK, it felt like a mixed blessing. He noted: 'It was common knowledge that the Chrysler multi-bank engine was full of bugs.'

Visiting the plant a couple of months earlier, another key customer, a Soviet engineer, heard the sales pitch for the new 30-cylinder petrol engine, relaying to Moscow what an American colonel at the plant had told him: 'he highlighted the advantages of the Chrysler engines, built based on automotive engines that the factory has built millions of. He mentioned that there will be no problems with supplies, and like car engines, they will be familiar and easy to service for any driver.'

Brigadier Ross accepted similar assurances, and would write later that the Detroit automakers would not let a substandard product leave the factory because, unlike those in Britain, 'the Americans were infinitely more jealous of their industrial reputation.' Deploying civilian teams of engineers to Egypt to talk to the customers and iron out faults, the American tank builders ensured that breakdowns became less frequent.

Ross's description of the supply chain is fascinating, giving a snapshot of the industrial challenge of tank building in the mid-1940s. Each Sherman M4A4 had 4,537 component parts: 1,269 were made by Chrysler themselves, 3,268 by subcontractors. The sourcing of these was carried out by company men, not a government procurement agency. At one point Chrysler was bringing in 450 new subcontractors a month, eventually having 8,079 of them in 856 cities.

'The Germans never learned to think in terms of reliability as we use the word,' claimed a Chrysler booklet published in 1946, 'that is, maximum performance with minimum care and replacement.' This bit of corporate grandstanding, titled 'Tanks Are Mighty Fine Things', argued furthermore that the Germans had 'a weakness for technical prowess at the expense of dependability'. Many Allied tank crews appreciated this reliability, but ended up feeling that there was something about those German machines, a raw menace, that their trusty Chryslers could never quite match.

With the American tank production behemoth, once it started churning out armour, it would prove difficult to change direction quickly in the light of new developments, such as the appearance of Germany's Tiger and Panther tanks. And the Sherman itself had been designed as part of an American doctrine or set of ideas about how war should be fought that was quite distinct from those of its British ally or indeed their common German enemy.

General Chaffee, placed in command of the re-formed tank force in 1940, set about forming armoured divisions as instruments of exploitation – they were to expand the breakthroughs won by infantry. With Roosevelt's decision to put America on a war footing almost one year before Japan's attack on Pearl Harbor brought the country into the global conflict, 1941 was a year of headlong mobilization.

Middle-ranking officers gained dizzying promotion, innumerable army camps sprang up in the heartlands, and great training exercises or manoeuvres played out across the US. It was a time when those with the right leadership skills, or indeed absence of self-doubt, were able to carve themselves great fiefs. Taking their lead from political and corporate America, many of these general officers courted publicity, sought to impose their personalities on their commands and tolerated little dissent. Patton, whose genius for war coexisted with a racist worldview and contempt for human frailty, was just one example. His image-building involved sporting a cavalry crop and breeches as well as pearl-handled revolvers.

Lieutenant General Lesley McNair emerged as another key character, particularly after Chaffee's death. Given command of the army's combat arms, he had the task of mobilizing dozens of new divisions as well as specifying their organization and doctrine. A graduate of the West Point class of 1904, 'Whitey' McNair (the nickname derived from his hair) had grown up in the Midwest and served in the artillery before being posted to the expeditionary corps in France in 1918, by which time he had reached the rank of brigadier general.

An adoring 1940 newspaper profile described the general as 'a gentleman who has silver in his hair, gold in his teeth, and lead in his pants'. The article's description of McNair as 'ruthless in making and breaking careers' gives a sense of how such qualities were positively viewed in the US Army back then. Given his considerable military experience, a marked contrast with so many who reached high rank at the time, his keen intellect and success at growing the army by more than 1 million soldiers in one year, it is not hard to understand why the army's boss, General George Marshall, rated McNair so highly. Marshall famously called him 'the brains of the Army'.

During the months following Hitler's blitzkrieg in France and the Low Countries, Marshall and McNair analysed the implications of that stunning victory for their new army. In June 1941, McNair told a conference of army leaders that finding the best way to defeat enemy tanks was 'the largest single question facing the Army today'. Whitey McNair came to the view that the best way of doing it was with anti-tank guns. He argued: 'It is questionable economy to employ a tank costing $35,000 to destroy another tank of substantially the same value, when the job could be done by a gun costing but a fraction of that amount. It seems more logical to employ our guns to counter the hostile armoured force while saving our own armoured forces for use against profitable targets to which they are invulnerable.'

So how was this different to the 'sword and shield' tactic that the Germans had used in France or indeed that the Red Army would adopt at Kursk? What could be more logical than using easily hidden guns on their wheeled carriages to ambush the enemy's armour, stalling his attack, while keeping your own tanks in reserve, ready for the counter-stroke?

In the months after that 1941 conference, hundreds of thousands of troops were deployed in the Carolinas for military manoeuvres. These gave a chance for the army to further refine its ideas on anti-tank warfare – or as unhappy tank men were to claim, to rig the whole exercise for the benefit of anti-tank units, and at the expense of armour.

McNair started building up an entire element of the military, later called the Tank Destroyer branch, with this specific mission in mind. The implication of this, when done on an American scale of mobilization, was the army at one point intending to form 220 Tank Destroyer battalions, equivalent to around one quarter of its frontline strength. In the event, fewer than half this number of units were formed, for reasons that will become apparent.

These battalions were meant to be used not just in defence, like the shield concept, but 'aggressively' too and in large numbers quite separate from tank units. And since the Tank Destroyer elements were either unarmoured, like the towed guns, or weakly so, they were meant to compensate by 'exploitation of their mobility and superior observation'. This was where American doctrine went off in its own peculiar direction, by arguing that these anti-tank elements, like clouds of skirmishers on the 19th-century battlefield, could lead the attack, as well as stiffening a defence.

The main armament initially intended for these units, 37mm anti-tank guns, soon became obsolete. Later they were replaced with 76mm towed guns and mobile 'gun motor carriages' armed with 75mm guns, including tracked vehicles like the M10. In the latter case the US Army, having started with McNair's injunction that anti-tank guns were far cheaper, would end up with a 76mm anti-tank weapon mounted in a turreted, tracked, armoured vehicle, having many components in common with the M4 Sherman and looking to any layman pretty much the same. So, given the scale of effort devoted to the Tank Destroyer project, what role were the actual tanks, the M4s, meant to be playing?

The 1941 Field Manual on operations noted: 'The armored division is organized primarily to perform missions that require great mobility and fire power . . . its primary role is in offensive operations against hostile rear areas.' In other words, it was to be committed for the exploitation of breakthroughs. McNair himself

4. The T-34 was ground-breaking in its protection, this view offering a sense of how both hull nd turret exploited sloping armour.

15. The Soviet Union's ability to keep producing large numbers of T-34s despite the overrunning of the Kharkov plant was a prodigious industrial achievement.

16. In its mature form, with a new turret armed with an impressive 85mm gun, seen here during the Korean War, the T-34 would last in service for decades.

17. It fell to the British, here the 8th Armoured Brigade in October 1942, to give the Sherman its baptism of fire, with some of the vehicles being taken from US Army units to strengthen the 8th Army before the Battle of El Alamein.

18. American production methods would produce huge numbers of tanks that were sufficiently reliable to win the affections of their crews.

19. With its uprated suspension and engine, as well as a turret redesigned to take a 105mm gun, the Sherman lived on in the Israeli army, with large numbers taking part in the 1967 war.

20. Wartime adaptation of the basic Sherman included fitting 'wading trunks' to many tanks involved in the D-Day landings. The extended air intake and exhaust allowed them to drop from landing ships into several feet of water, though in this case the sand was too soft to bear its weight.

1. The '17-Pounder Tank' or Firefly as it later became known, fitted a British-made anti-tank gun capable of knocking out Tigers and Panthers to a modified Sherman turret.

22. A command variant of the Tiger belonging to the 503rd Heavy Tank Battalion on the Eastern Front during the winter of 1943–44 when small numbers of these tanks were used to blunt large-scale Red Army attacks.

23. Inside the plant at Kassel where Tigers were built, peaking at around 100 per month in May and June 1944. It was construction on a small-scale, high-cost basis that could never outweigh the prodigies of US and Soviet manufacturing.

24. The world's only operational Tiger, the famous 131 that was captured in Tunisia and now resides at The Tank Museum in Dorset.

put it succinctly: 'our general concept of an armored force [is] that it is an instrument of exploitation, not greatly different in principle from horse cavalry of old.'

Like many a doctrinal flight of fancy, for example the British distinction between infantry- and cruiser-type tanks, McNair's concept required the cooperation of the enemy, something generally in short supply on the battlefield. If an anti-tank gun or open-topped tank destroyer could be swiftly dealt with by airburst shells (unlike a tank), then that's what the enemy often did to them.

There were a number of consequences of the US Army adopting these ideas. Lives and careers were lost, of course, by men keen to demonstrate the doctrine was sound. And tank destroyers, because of their mission, were given priority over tank units when guns or ammunition with superior armour-piercing capabilities were fielded.

Inevitably, perhaps, the mismatch between these ideas and battlefield reality was not long in coming, but the consequences for the Sherman and those crewing it could not be so quickly remedied because the US Army had opted for McNair's ideas on such a grand scale.

By dawn on the 14th of February 1943, it was clear it would become a day of hard learnings for the US Army. Rommel, dubbed the 'Desert Fox' by the British press, was being hemmed in by advancing Allied armies, and, having seen the supply difficulties facing Montgomery's 8th Army, the German general turned his attention to those approaching from the west.

American ground forces had entered the North African war in earnest after landings in Morocco and Algeria three months earlier. As the Axis forces attacked through the rugged terrain of the Tunisian/Algerian border lands, thousands of US troops from the 1st Armored and 1st Infantry divisions were to receive their baptism of fire.

Exploiting a sandstorm to cover their preparations, a German *Kampfgruppe*, a combined arms fighting force with 200 tanks and thousands of men, had attacked through Faïd, a small town held by the Americans. Realizing the Germans had penetrated their positions and were bypassing resistance, US commanders sent forward a battalion of 47 tanks. With the poor weather lifting, they were bombed from the air before being engaged by German tanks, meaning that by mid-afternoon 40 of those tanks had been knocked out. American forces were falling back towards a key choke point through the *jebel*, the Kasserine Pass.

The following day, to buy time as reinforcements moved up and their commanders tried to pull together the remnants of the fighting on the 14th, they launched a counter-attack. Around midday on the 15th of February, another tank battalion was sent forward, with tank destroyers on its flanks. The tank destroyer battalion engaged in this fight had 37mm and 75mm guns mounted in M3 halftracks. It was a sub-par arrangement in which even those with the heavier guns had no advantage, firepower-wise, over the Lee and Sherman tanks they accompanied, but suffered from poorer mobility and far worse armoured protection.

Axis forces soon took apart this counter-attack with artillery fire (to which the open-topped tank destroyers were particularly vulnerable), as well as their own anti-tank gun and tank fire. Burning American armour soon dotted the Tunisian scrub, and by 6 p.m. the attack was broken off. In two days' fighting the 1st Armored Division had suffered the loss of 98 tanks, 57 halftracks (of various kinds) and 26 artillery pieces. They were also forced back 50 miles. During the following days, US forces (the British and French were also engaged in this area) would return to Faïd and the other positions they'd been driven from. But as a consequence of this costly setback the Americans had to send up 7,000 troops and nearly 200 tanks to replace their losses.

Recriminations soon started, with various commanders relieved of their posts. In the context of our story, there were arguments about tactics and whether the top brass's idea of confronting panzers with 'aggressive' tank destroyer units made any sense at all.

Reports sent back to McNair and his boss General George

Marshall were scathing about the obsolescence of the 37mm gun. One of Marshall's staff in Tunisia was bluntly heretical in questioning the whole concept: 'the tank destroyer has in my opinion failed to prove its usefulness.' Major General Jacob Devers, head of the Armored Force, so evidently a man with a vested interest, joined the bandwagon, reporting that the 'tank destroyer arm is not a practical concept on the battlefield'.

Like many an ideologue confronting failure, McNair doubled down. Writing to the boss of the tank destroyer training school two months after Kasserine, the general argued: 'the tactics taught at the Tank Destroyer School are not applicable to this theatre.' What was to be done? Battle drills would have to be rethought in order to prove the soundness of the concept; then McNair added, 'I trust you will not allow these developments to destroy the aggressive spirit.'

In practice the early setbacks did mean modifying the concept: less of the charging about solo originally planned and instead closer integration of tank destroyers into the wider plan (much like other armies, in other words). But from the point of view of America's mighty industrial and troop-training effort, tank destroyers continued to be what the likes of Devers and Patton thought was a diversion of effort from the tank force.

Building the M10 tank destroyer, for example, literally took away engines, bogies and tracks that could have been used to construct more Shermans. Fitting M10s with 3-inch guns allowed the Army Ground Force HQ to argue it was taking steps to answer the deployment of heavier German tanks while dragging its heels over up-gunning the Sherman. For in the aftermath of Rommel's offensive, early in 1943, the Tiger began to loom large in British military minds, even if it took the Americans a little longer to see its importance.

Following the Allied landings in November 1942, and the introduction of substantial American forces into the North African theatre, Hitler had ordered reinforcements on the Axis side. Additional Luftwaffe squadrons and paratrooper battalions were the first to fly over, followed by the shipping of further divisions and two battalions of Tiger tanks. Although there were fewer than 50 of these behemoths, their appearance worried Allied tank crews greatly. The story of this machine will be told in detail in the next chapter, but its firepower and armoured protection were both much better than those of the Sherman. Around 20 Tigers had been used during the battles of February–March 1943 in western Tunisia. Seven had been lost, not due to Allied action but having been destroyed following breakdowns by the Germans prior to falling back.

On the 24th of April, on the other side of Tunisia, where the 8th Army was pushing up from the south-east, British troops succeeded in capturing a Tiger. This vehicle, with the tactical number 131, is today at The Tank Museum in Bovington, and is the only working example in the world. Tiger 131 had been involved, that day in 1943, in an attack on a hilltop held by the Sherwood Foresters. Having initially bombarded those trenches from a safe distance, the Tiger platoon tried to overrun them. Advancing up the hillside, they were engaged by Churchill tanks from a neighbouring height. Traversing its turret towards this new threat, Tiger 131 fired its 88mm gun, knocking out one of the Churchills.

The infantry had been given a brand-new shoulder-launched weapon, the Projector, Infantry, Anti-Tank or PIAT. Firing the anti-tank bomb at the Tiger, they were disappointed to see it bounce off without having any effect. But then one of the Churchills, armed with a 6-pounder (57mm) gun, scored what was either a skilful shot or a very lucky one. It lodged just where the panzer's turret joined the hull, jamming the traversing mechanism and making the vehicle very vulnerable. The crew bailed out and were captured.

As the war in North Africa drew to a close, the Sherman's period of popularity with its operators had reached a peak. Although they had heard chilling tales of the Tiger, few had encountered it. Rather, in the 8th Army at least, experienced crews who had fought their way across North Africa to the gates of Tunis, the last

Axis stronghold, during 18 months since Alamein had high confidence in these machines. Their tactics were slick too.

On the 7th of May, for example, not far from the Tunisian capital, troops of the 7th Armoured Division, the famed Desert Rats, encountered defences that might quickly have decimated an armoured squadron a couple of years earlier. Men from the 5th Royal Tank Regiment (RTR) detected a battery of 88mm Flak guns deploying atop a ridge. There were at least four 88s, and one of the 50mm anti-tank guns.

Moving gingerly into fire positions on a neighbouring ridge, they kept their Shermans 'hull down' so just the turret was unmasked to the enemy, a much smaller target, and started firing on the German gun positions. Lieutenant George Thomson, one of the 5th RTR troop commanders, was subsequently awarded the Military Cross, the citation reading: 'He soon knocked out two of the 88mms and the 50mm. The other two 88mms were so accurately engaged by him that he forced the crews to abandon the guns, which were later captured.' The 88s had opened up, of course. But none of the British tanks were hit. The 75mm gun, lobbing its shrapnel shells, had made a critical difference. Their explosive power dealt with enemy crewmen and their range nullified the Flak gun's advantage over earlier models of British tank.

As this action occurred, off to 5th RTR's right, a reconnaissance patrol from sister battalion 1st RTR had come under fire from one of the few remaining Tigers. The force of the 88mm shell striking the rear of one officer's Crusader tank was so violent it spun the tank around and stopped it dead. This time the tank guns turned on the panzer did not decide the issue. Instead, with 1st RTR's advance temporarily stalled, an artillery bombardment was summoned up, and dozens of shells falling about the Tiger caused it to quit its hilltop position. The North African campaign was concluded soon afterwards.

British tank crews, particularly those of the 8th Army with their long experience of woeful homespun tank designs, ended the desert campaign with great affection for the Sherman. Never mind the bracing experience that some had with Tigers; its armament was capable of dealing with Panzer IIIs and IVs, neutralizing enemy anti-tank guns at distance and supporting infantry in the assault. Its armour was perfectly adequate, and automotive reliability won Sherman drivers' loyalty.

Early users in late 1942 and 1943 had given their feedback to the manufacturers' representatives in Egypt, and it had also made its way back to Brigadier MacLeod Ross, the British army representative in the United States, whose job was to liaise with the makers concerning future shipments. Brigadier Ross wrote that what they wanted was: the diesel engine version, better ammunition stowage to reduce the chance of catastrophic hits, and wider tracks to improve mobility over soft ground. In this last point, and preference for diesels, British experience was similar to the Red Army's. The T-34 remained a better bet over the Ukrainian mud because its lighter weight and wider tracks gave lower ground pressure, as did the design of the track links themselves. In time, the Americans developed 'grousers', attachments that could be clipped to the caterpillars' outside edge to widen them.

The issue of ammunition stowage was also dealt with – over time. The likelihood of penetrating rounds setting off shells led to the Sherman gaining the grisly reputation for catching fire, leading to the nicknames 'Ronson' (after the lighters) and 'Tommy Cookers' from the Germans. Eventually 'wet stowage' Shermans were introduced, with the 75mm ammunition stowed lower down in the hull and jacketed in lockers containing a fire-suppressant liquid.

Ross would argue that it was the question of firepower that emerged early in 1943, following the appearance of Tigers, that was never satisfactorily addressed by the Americans during the war. In fairness, the British War Office took time to get it right too. Crews returning from North Africa and Italy were horrified to be given the Cromwell, a new British tank with the same armament as the Sherman and slab-sided armour. Monty's message of November 1942 that 75mm-gun tanks were 'all that was needed' had fed through to Whitehall and the boffins in research establishments. So that's what the Cromwell had been given. In time, its operators learned that the vehicle's Rolls-Royce Meteor engine was much more reliable than those of the earlier British cruisers, giving it a greater turn of speed than the Sherman.

What the British did succeed in doing, and the US Army did not, was finding a good way of boosting their tank battalion firepower in time for the D-Day landings in June 1944. For while the UK's wartime record in tank design was middling to say the least, the section of the Ordnance that developed artillery was very well run. The Royal Artillery boss of this organization, Major General Campbell Clarke, oversaw the development of two outstanding anti-tank guns during the war: the 6-pounder (or 57mm gun) and the 17-pounder (a 76mm weapon). On a small, wheeled carriage the first of these was deployed by the hundreds in the anti-tank batteries of infantry battalions. It was also installed in Crusader and Churchill tanks, having been the weapon that disabled Tiger 131.

Clarke was a scientist, almost 58 when these questions came to a head in 1943, having served all but 20 of his years in army roles dealing with the production of ordnance at the historic centre of artillery development, Woolwich Arsenal. From the start of the war he had waged a 'campaign to design tanks for weapons and not vice versa'. It took two years for him to persuade the army to fit the 6-pounder, developed at the start of the war, to tanks. The success of that weapon (in an armour-piercing role) compared to guns firing bigger shells (like the US 75mm or 76mm) owed much to the speed at which its shot travelled. So even though its armour-piercing shot weighed just 3.24kg, the 6-pounder fired it at a muzzle velocity of 831 metres per second.

Clarke's team did not rest on their laurels. Reviewing the intelligence received in April 1943 about the inability of the Sherman's 75mm gun to knock out the Tiger, he noted: 'all possible steps must be taken to increase the punching power of the 6-pdr or of a more powerful successor to it.' So they developed a new round for the 6-pounder, and an entirely new anti-tank gun, the 17-pounder (or 76mm). In boosting the penetrative power of the 6-pounder, they relied on an ingenious principle evident since the days of knights in armour. A weapon in which force is concentrated on a small point is more likely to punch through – it's the same principle by which a stiletto heel might damage a dancefloor but a more conventional shoe would not.

The new 6-pounder shell, once fired, shed a metal casing (or sabot) wrapped around it, continuing on its way with all the energy imparted from the initial blast but striking the target with a smaller point. This Armour Piercing Discarding Sabot (APDS) shot flew at twice the muzzle velocity of the Sherman's 75mm one, which more than compensated for the fact that the part hitting the enemy tank weighed just 1.42kg. Even before the development of that new APDS shot, the US Army, seeing the woeful performance of its 37mm anti-tank guns in Tunisia, had adopted the British 6-pounder. It was manufactured on a suitably American scale as the 57mm gun.

When it came to Major General Clarke's next invention, the 17-pounder anti-tank gun, British and US armies went their separate ways. The new weapon had appeared on wheeled carriages in October 1943, during the final months of the desert campaign, drawing great plaudits from its gunners. It achieved a similar muzzle velocity to the original 6-pounder, but fired an armour-piercing shot of much greater mass, 7.7kg. Setting all the stats to one side, the critical thing was that the 17-pounder could penetrate the Tiger's armour and would be able, likewise, to knock out the new Panther from the front, once that appeared.

Instead of buying into the new British product as they had its predecessor, the US Army chose a homegrown 76mm gun of inferior performance. Ross claimed the US gun gave 'inadequate performance, which was never improved by higher velocity ammunition', and later put the American decision down to 'false pride'. When the American generals wanted something more capable of matching up to the Tiger's 88, they adapted a 90mm anti-aircraft gun for the purpose. But this would not fit in a Sherman turret, and when a new one was designed to accommodate it, priority was given to the Tank Destroyer force.

So this tangle of national pride and doctrinal stubbornness would lead to thousands of Sherman tanks being deployed on the Normandy beaches in June 1944 with the same 75mm gun that had already shown its limitations in Tunisia more than one year earlier. US Army Shermans with the new 76mm gun did begin arriving several weeks later.

For the British and Canadians, though, an answer had presented itself well before D-Day. As early as April 1943 Major General Clarke, the owlish Director General of Armament at Woolwich, started advocating adapting the 17-pounder for use in a vehicle. Clinging to the doctrine formulated the previous November that the 75mm gun was 'all they needed', the General Staff initially fought a rearguard action against this eminently sensible idea. Such was the poisonous atmosphere in the War Office committees that Clarke lobbied the Minister of Supply to make the argument on his behalf, because 'such a suggestion on my part might well create resentment and produce more harm than good.'

By September 1943 the battle was won, with work starting to fit the 17-pounder to a trial Sherman. It was a beast of a weapon. Newsreel footage of these anti-tank guns firing shows the kicking recoil that came from an explosive charge large enough to achieve its formidable muzzle velocity. Add to this that the 17-pounder breech was too big to fit in the Sherman turret, its shells too large to be carried in the same racks and its firing cycle had not been designed for an enclosed space, and it gives some idea of the obstacles that had to be overcome. This last quality meant that when the breech block dropped, a fraction of a second after firing, and the spent casing ejected, there was still burning propellant inside.

British designers came up with a solution called, depending on the version, the Sherman IC or VC. Generally they were referred to as '17-pounder tanks' but later on they picked up the name Firefly. The 17-pounder was turned so its recoil mechanism would fit, and the back of the turret cut off so an armoured extension could be welded on. This allowed radios and other equipment to be moved further back so they weren't smashed by the 17-pounder breech as it flew back into the turret interior when firing.

Changes were made to the hull too. The Sherman's fifth crewman, sitting next to the driver, operating the hull machine gun, lost his place in order to make room for the gun's bigger shells. As for the burning propellant ejected into the confines of the turret, it was felt, with typical wartime sangfroid, that it was just a small proportion of the explosive charge, and the crew could put up with it.

Firefly crewmen, trying the tank out on the ranges, were taken aback by the power of the 17-pounder. The kick of its recoil, as well as the flash and dust from its muzzle blast, soon brought the realization that smart tactics would have to be used because of the risk that an enemy might quickly spot them. They also discovered, removing their berets after the training shoot, that a tan line had appeared on their foreheads: they had been lightly grilled, a consequence of the smouldering propellant ejected into their turret space.

Gerry Solomon, a sergeant in the 5th Royal Tank Regiment, told me in an interview in 2014 that he had been sent up with other crews to the factory in Nottingham where their Fireflies had been built to collect them just six weeks before D-Day. The works manager told him he was to command 'a secret weapon', much to Gerry's amusement.

Production of the Firefly was sufficient by this point to give leading armoured units one per troop (a unit of four tanks), thus each one arrived in Normandy in June 1944 with a single 17-pounder-armed tank and three conventional Shermans or Cromwells. It wasn't perfect – the old question reared itself of how to make sure the right vehicle was in the right place when a Tiger or Panther appeared – but it made a tremendous difference.

Lieutenant General Omar Bradley, commanding the American 12th Army Group in this battle, would later remark: 'only the

British had found a weapon to pierce the Panther's thick-skinned front in their tough old [*sic*] 17 pounders . . . when I queried Monty to ask if the British could re-equip one Sherman in each US tank platoon, he reported that ordnance in England was overloaded on British orders.' The British had got their up-gunned tanks into action remarkably quickly, but there were limits to their industrial capacity.

By the 7th of August 1944, German defences in Normandy were collapsing. One week earlier General Patton had been given the newly created Third US Army as the Americans launched Operation Cobra, breaking through the defensive lines that had held the Allies at bay since D-Day.

Driving in an open jeep with one of his staff officers along a French road that was a scene of carnage, broken German bodies and vehicles lining the sides, Patton exulted: 'Just look at that, Codman, could anything be more magnificent?' The general reportedly believed that leaving enemy bodies out, while quickly recovering those of his own units, was good for morale.

Certainly, the pace his troops had achieved in those few days was a tonic. The Third Army consisted of 200,000 troops organized in eight infantry and two armoured divisions. Its infantry had plenty of trucks and jeeps, contributing to the army's total of 40,000 vehicles, so they had no trouble keeping up. Patton's wanton aggression had been expressed in a speech at headquarters shortly before he hurled his army into action: 'Forget this goddamn business of worrying about our flanks . . . we don't want any of that in the Third Army. Flanks are something for the enemy to worry about, not us . . . we are advancing constantly and we're not interested in holding on to anything except the enemy. We're going to hold on to him by the nose and we're going to kick him in the ass.'

Once through the gap smashed in the German defences at Avranches in western Normandy, Patton's divisions drove hard, fanning out to the west and south, and eventually turning east as part of a move to work around behind the German divisions pinned frontally by the British and other Allies and to sew them up in what became known as the 'Falaise pocket'.

The enemy, realizing the danger posed by the American advance, had indeed tried to threaten Patton's eastern flank by launching a counter-offensive, Operation Lüttich, on the 6th of August. This drive from east to west by elements of four panzer divisions was meant to cut off American units streaming south and to stabilize the German position. It produced large-scale fighting around the town of Mortain when *Kampfgruppen* equipped with around 250 panzers and 60 assault guns were sent into action.

As Patton admired the destructive aftermath on the road near his headquarters, furious battles raged in French villages to his rear. The German columns, spearheaded by Panthers, were met by US infantry backed with anti-tank guns. Indeed, in many places the guns copied from the British 6-pounder design struck their targets, setting fire to them or 'brewing them up' in army parlance. The Allied gunners had learned that this weapon could readily punch through the Panther's side and rear armour.

Later on the 7th of August, some companies from the 3rd Armored Division were thrown into the fight, with both sides' armour taking losses. Air attacks, from P-47 Thunderbolts, added to the panzer divisions' difficulties. The American infantry held the line, and dozens of Panthers were knocked out. Such were the debates between the different elements of the American military about who had really defeated these attacks that an operational analysis team was sent into the field to inspect the wrecks and report back.

Surveying 127 destroyed German vehicles (of various types) in one sector, they attributed 33 to air attack versus 35 by direct action by ground forces – though they had trouble attributing cause in 41 cases. When it came to 33 Panthers they inspected, 14 had been knocked out by direct ground action (it was not possible to differentiate between hits by tank destroyer or tank platoons), six by air strike, and ten had been abandoned. This last figure highlighted the degree to which a significant number of panzers that were on paper superior were lost due to mechanical breakdowns or fuel shortages.

Patton's gas-guzzling armoured divisions, on the other hand, set the pace. And in addition to all their wheeled transport, each of his infantry divisions had its own tank and tank destroyer battalions. Battle experience had taught the US Army to integrate its anti-tank weapons closely and to deploy them, as the Germans had in May 1940, in the shield role when danger threatened.

Third Army's troops continued to forge south and to turn eastwards, exploiting their success. Germany's position started to collapse and Patton's army raced on, crossing the River Seine and on towards the Meuse by early September, that barrier on France's eastern borders whose crossing by German forces in May 1940 had such dramatic effect. Patton's troops had driven hundreds of miles, liberating a swathe of that country, an achievement he described with characteristic braggadocio: 'the Third Army has advanced farther and faster than any army in the history of war.'

The Sherman tank played a key part in this, as a potent fighting machine with reliability built in by the American motor industry. The manoeuvre conducted by Patton in August 1944 fitted very neatly the template he had set out 26 years before in one of the US Army's first papers on the purpose of armour: 'If resistance is broken and the line pierced, the tank must and will assume the role of pursuit cavalry.'

This idea, of armoured divisions being there to exploit success won by their infantry brethren, had been voiced over and over again in the doctrines of Whitey McNair and other American generals. The Sherman was designed to perform that role. It had also, as tank battalions were integrated into the infantry divisions, to act as a fire-support weapon for them, like German or Soviet assault guns. When German panzers threatened to break through, the tank destroyers were meant to stop them. And certainly, with the failure of Germany's counter-attack, Operation Lüttich, in which easily concealed anti-tank guns played a notable role, you could say that the US Army 'system' worked.

Having placed its bet on the Sherman design by 1941, the American military improved it incrementally – but did not manage to field a successor, the T26 Pershing, until February 1945. Instead, they produced very large numbers and highly reliable tanks. Another of McNair's doctrinal principles was to allow the user to drive improvements in armoured design, rather than staff back in the US doing it, a fine idea in principle but one that ignored the time lag between a problem appearing in the field and the time required to develop a solution. This was all well and good on the level of national doctrine, rather less so if you were a Sherman crew facing a Tiger or Panther in the Normandy hedgerows in June or July 1944. British or Canadian units had an answer, at least a partial one, with their Fireflies, as we will see in the next chapter, but many a US Army crew went to their makers because they were overmatched.

While American soldiers sent to war in a Sherman often complained bitterly about its shortcomings, feeling it was obsolete by the summer of 1944, the odds of them coming face to face with a Tiger were not at all high (in Normandy, indeed, they were almost zero). It was a matter of happenstance that the Americans did not experience large-scale tank battles in Europe until after the Normandy campaign; even then for much of the war in this theatre the Sherman remained an infantry support weapon, with loadouts of 90 per cent high-explosive shells being common.

So what was McNair's verdict? Did he argue that Patton's breakout, and the failure of the German counter-attack to stop it, proved he was right about both armour and the best way to counter it (with cheap anti-tank guns rather than other tanks)? Sadly, he did not live to see it. In July 1944, as Operation Cobra was launched, the general went to observe a vast aerial bombardment that marked its start. He was one of scores accidentally killed by stray American bombs.

However, writing in 1945 shortly after these events, the chief of staff at Army Ground Forces claimed a posthumous vindication on his old boss's behalf. Major General James Christiansen admitted that there had been times in Normandy when armoured companies ran into great difficulty, but added: 'Our tankmen there benefitted by the superior mobility, mechanical dependability and accurate fire of the Sherman.' General McNair, Christiansen noted, had never pushed for a gun with 'punch'.

After 1945 the Sherman was exported to countries worldwide. That lack of firepower was addressed in different ways, the Israelis for example rebuilding turrets to take 105mm guns. From Korea to Vietnam, the Sinai to Pakistan, Shermans served on for decades.

Inevitably, though, there are many, looking at the history of the Second World War, who argue that the standout tank of the conflict was the one that has stalked both this and the previous chapter. It was the vehicle that Sherman and T-34 crews most feared coming face to face with, and which scored some of the greatest successes of that bitter conflict.

Panzerkampfwagen VI, Tiger

Entered service: 1942
Number produced: 1,347
Weight: 57 tonnes
Crew: 5
Main armament: 88mm gun
Cost: 300,000 Reichsmarks (approx. £30,000 in 1942, which is £1.35m in 2024 pounds)

Tiger

Things began very badly for the 162nd Grenadier Regiment that morning, the 17th of March 1944. They had first been hammered by an earth-shattering Red Army artillery bombardment. Then the tanks had rolled over their trenches, followed up by infantry. German soldiers were streaming back from the forward positions near Narva (in modern Estonia); dozens, having dropped their weapons, raced across the snow-covered fields. Their forward line had gone and the enemy was heading for the main defensive belt, running east–west along a road.

It was at this moment that Otto Carius, a 21-year-old junior officer in the Wehrmacht, acted to stop them. He was in command of two Tiger tanks that began moving out of hide positions and along the road, at right angles to the Soviet attack. As the drivers opened their throttles, with the huge vehicles picking up speed, Carius realized they were coming under fire from their right flank. The enemy had moved anti-tank guns onto a railway embankment, and their shells began striking the Tiger's side. Panning the turret in their direction, Carius immediately issued orders to engage them, and the big 88mm gun began its thunderous fire.

Having knocked out four or five anti-tank guns he resumed his advance, and was startled to see that 'five T-34s were already closing at full speed on the *Rollbahn* [the road forming their main defence line].' The enemy hadn't just sped over the railway embankment; they now threatened the main defence line. 'It was literally a question of seconds,' he would later write. 'We were lucky that the Russians had buttoned up [i.e. closed their hatches] like they always did and could not size up the terrain fast enough.' Another T-34 was just metres away from the other Tiger under his command. Quickly Carius gave a warning over the radio: 'Kerscher took care of the Russians with a direct hit. They careened into a bomb crater and didn't come out . . . the remaining five T-34s didn't even get to fire. They probably also didn't have a clue as to who knocked them out and from where.'

They had stemmed the immediate crisis. But the Soviet attack was continuing, with fresh waves being committed. Carius and Kerscher were able to push forward in their Tigers so they were facing the onslaught head-on, as well as using the railway embankment as cover. That afternoon they knocked out another five T-34s.

A running battle continued along this line for days. Having accounted for 14 T-34s and one KV heavy tank on the 17th, they killed another four on the 18th and nine Russian tanks on the 19th. During two weeks of fighting it would be claimed in the Nazi press that Lieutenant Carius had knocked out 26 enemy tanks. An army newspaper went with a paean of praise to the men and their fighting machines: 'Whenever the Bolshevists arose out of the woods or stormed across meadows or through swamps, the Tigers shredded the waves of earth-brown attackers with high-explosive rounds. Whenever the masses of Soviet tanks rolled out of depressions and hollows towards our positions, then the cannons of the Tigers shattered the steel colossus.'

There was no doubt that in tank-to-tank terms it was a grossly unequal contest. Carius had been wounded during the early weeks of Operation Barbarossa when a T-34 had knocked out the little Czech-built Panzer 38(t) he was crewing. But now there was no comparison: the Tiger's frontal armour was 100mm, double that of the Czech tank. Its overall weight was 57 tonnes as opposed to 9.8 tonnes. Most importantly, the Tiger's 88mm gun could regularly knock out T-34s at 1,000 metres (penetrating 100mm of armour, angled at 30 degrees, at that range in Wehrmacht tests), whereas the T-34 had virtually no chance of taking out the German machine except with a flank shot at 500 metres or less.

Carius, describing that fight of March 1944, was in no doubt that the T-34's poor turret design and Soviet orders to fight closed down compounded the unequal nature of it: 'Tank commanders who slam their hatches shut at the beginning of an attack and don't open them again until the objective has been reached are useless.'

Having made its debut on the Eastern Front late in 1942, the Tiger had become the top-tier predator. Its mere appearance could cause panicked Soviet tank crews to abandon their vehicles. Nikolai Zheleznov, a Russian tank commander, noted after the war that, as a Tiger panned its gun towards you, it might be possible to get a shot off, but once it started to adjust the elevation of its gun it was time to bail: 'you'd better jump out or you could get burned.'

As we saw during the Prokhorovka battle of the Kursk offensive, Soviet generals had tried to explain their setbacks by claiming that fantastic numbers of Tigers had been there, and to have knocked out 55 of the beasts, when they actually only accounted for one. Naturally, the gross inequality of these contests appealed to Nazi propagandists, who extolled the feats of their panzer *Übermenschen* or supermen, frequently comparing them to the Teutonic chivalric warriors of old. 'As once the knight with his armed steed must have been unified into one fighting entity,' the newspaper piece on Carius declared, 'so are the Panzertruppen one with their steel fighting vehicle.'

SS officer Michael Wittmann, whose Tiger troop was in action at Prokhorovka, received similar glowing write-ups. He and Carius received the knight's cross award, Wittmann's citation in January 1944 claiming he had notched up 66 kills.

With both men and machines, this idea of German excellence defeating superior numbers, the 'Red hordes', appealed to Hitler personally and was thus communicated throughout the army and industry. Indeed, in May 1941, shortly before series production of the Tiger was ordered, the Führer declared at a conference of tank designers: 'technical superiority is of decisive significance . . . even small series of superior weapons can be decisive.'

This philosophy spawned the Tiger and several other heavy armoured vehicles as the war went on. In terms of that eternal triangle of armoured vehicle design, Hitler believed in the constant boosting of firepower and armoured protection, while scorning increased mobility. 'The faster ship has only one advantage: to utilize its greater speed for retreating,' he told his armaments minister. 'It's exactly the same for tanks.'

When it came to demonstrating the superiority of German crews and armour through grossly disproportionate kill counts, the Tiger became the weapon of choice. Soviet doctrine meanwhile tried to discourage tank commanders from getting 'distracted' by enemy armour, focusing instead on supporting the infantry. Though the Soviet military would also develop the IS-2 heavy tank, which could go muzzle-to-muzzle with the Tiger, it still emphasized the importance of anti-tank guns in destroying German armour.

In April 1943, as the British debated what to do about the new German heavy, Winston Churchill minuted his 'Most Secret' thoughts to the Defence Committee (Supply). He suggested that British tanks of 60, 70 or even 80 tons might be required to meet the threat, arguing with a typically Churchillian flourish: 'The warthog must play his part as well as the gazelle.' But British industry in 1943 was quite unable to come up with a tank of that class, lacking an engine powerful enough.

The US Army meanwhile had, several months after encountering the Tiger in Tunisia, rejected a proposal to build an equivalent (the T26), and in doing so dismissed the entire concept behind the German vehicle. Lieutenant General Lesley McNair's HQ told the army chief in November 1943: 'There can be no basis for the T26 tank other than the conception of a tank-vs-tank duel which is believed to be both unsound and unnecessary . . . battle experience has demonstrated that the anti-tank gun in suitable numbers and disposed properly is the master of the tank.' While this view changed late in the war, in time for a small number of T26s to reach Europe, American decisions in 1943 ensured that, following D-Day, US and Allied tankers would have to face the Tiger in their Shermans and hope that they would receive support from specialist anti-tank gun batteries.

In the German case, plenty of emphasis was given to combined arms operations, supporting the infantry, and so on, if you were commanding a Panzer IV or even a Panther in a standard tank battalion. But different rules were applied to the Tiger soon after the first heavy tank battalions were fielded. These were to be a vital reserve, used to turn the tide. As the commander of the 501st Heavy Tank Battalion reported from Tunisia, 'a concentrated commitment using Tigers at the point of main effort will result in success; this is due to their strong armour and powerful gun, and also their effect on the morale of enemy troops.' This psychological effect could be best achieved, he noted, by exploiting their superiority over enemy armour: 'Do not forget: combat against tanks is the main duty for the Tiger.'

Carius, Wittmann and the other commanders passing through the Tiger training battalion in Paderborn had been inculcated with this doctrine. This Westphalian town was sufficiently far from the front to rank as a restorative posting for the instructors. New arrivals were soon reminded of their chosen status as Tiger crews. Having passed through the special training, Carius wrote that his main mission was destroying tanks and anti-tank guns, and that 'the psychological support of infantry during covering missions is of only secondary importance.'

The elitism surrounding the heavy tank force came at a military and industrial price, as we will see. And even the Tiger's design, in creating a sort of muscle-bound prizefighter, had many flaws. But it undoubtedly became celebrated by friend, feared by foe, and was profoundly influential in the way armour, particularly in the West, developed after 1945.

Those who had lived in awe of Tigers in Normandy or the Ardennes would go on to create their own versions of them, feeling that Nato, the Western alliance, would need something similar if it was to hold back the vastly superior numbers of the Cold War Soviet armoured divisions. They too placed their faith in technological superiority over quantity.

The British Chieftain tank that I commanded had a lot in

common with the Tiger: heavily protected, of similar weight and with an extremely powerful gun. But it also shared the Tiger's faults, being overly sophisticated, underpowered, mechanically unreliable and so heavy and slow that it had to be used with great care. Thus the story of Germany and arguably the world's most celebrated tank is one that from the outset involved difficult compromises from those who built and used them.

The Tiger's birthplace was a railway engineering works at Kassel in the province of Hesse. Lodged between the switching yards, the town centre and the Fulda river was the Henschel Werk Mittelfeld. Across the way from its machine halls was another line where railway locomotives continued to be assembled throughout the war. But it was in the central plant on that site that Tigers were built, even if one has to qualify the notion that Germany's most feared wartime tank originated there.

As with the Panzer IV, the concepts underpinning its design originated with the panzer troops, while the Heereswaffenamt (or HWA) drew up specifications and managed projects. The story is further complicated by the fact that Krupp developed the Tiger's turret. But it was to Henschel that the creation of its hull fell.

Dr Erwin Aders was the firm's bespectacled chief designer, and, more than any other individual, he drew together and integrated the elements that would create the Tiger as a whole. Aders had spent a lifetime in heavy engineering and was already in his sixties when development work on the project got underway in earnest at Kassel.

Ideas about a heavy or breakthrough tank had been circulating in the Wehrmacht since 1935. It took six years of gestation before the designs and prototypes began to resemble the vehicle we now know. Throughout this time the debates eddied back and forth about the shape it should take. In essence the story was one of repeated demands for increased armoured protection leading to a heavier and heavier vehicle, which in turn prompted the need for all manner of engineering mitigations in order to retain some degree of mobility. It can be argued that these dilemmas were never fully resolved.

Thus, one of the early hull prototypes produced in 1940 by Henschel was designated VK30.01, a later one VK36.01, and a rival proposition from Ferdinand Porsche's eponymous engineering firm the VK45.01. In each case the first two digits revealed the weight in tonnes. At the time of the 30-tonne prototype, a 600hp engine was specified, giving a very respectable 20hp/tonne ratio; however, little by little this was eroded as the machine grew and grew.

In the wake of the 1940 campaign, with its unpleasant surprises about how poorly protected German tanks were compared to French heavies, the initial spec for a 30-tonne vehicle was quickly superseded. Yet the HWA continued to argue that the vehicle could not get any heavier, because of the weight-bearing limits of most bridges.

Adding to the uncertainties about the heavy tank design, the HWA pursued multiple prototypes from different manufacturers that embraced various technologies. So, there were turret designs using a long 75mm gun as well as the 88mm one that was eventually adopted. Meanwhile, Ferdinand Porsche used his friendship with Hitler to get a rival hull commissioned. The Porsche offer used an air-cooled engine and an experimental electric transmission.

In May 1941 the dilemmas concerning the turret at least were resolved by the decision to arm the Panzer VI, as the vehicle was called before the name Tiger was adopted, with the formidable 88mm gun. But Hitler's personal interference had started to muddle the industrial decision-making, and it was decided to order 100 hulls from Porsche as well as developing Henschel's one, leading to a competitive trial in April 1942.

'After working day and night,' Henschel's chief designer later wrote, 'the first operational vehicle was loaded with cross-country tracks onto a railcar and sent to Hitler's headquarters.' Porsche's offer, the Tiger (P), was likewise dispatched, and the rival vehicles put through their paces under the Führer's watchful eye. For weeks, the prototypes churned up the Prussian mud, but very quickly it became apparent that Henschel's was superior.

One of the top officials present described the Porsche prototype's performance as 'a total catastrophe'. There were repeated breakdowns, and one of the Porsche engines seized after just 60 miles despite consuming 85 litres of oil. The electric transmission also proved highly temperamental. Furthermore, the Alkett plant on the outskirts of Berlin was way behind in its contract to deliver Porsche's 100 hulls. Even having Hitler's favour couldn't save Porsche.

By June 1942, Henschel had made its first production vehicles, the HWA accepting them for army service two months later. How, then, had Aders and the other designers reconciled the many demands made upon them by soldiers, bureaucrats and politicians?

The 88mm gun, firing a 10.2kg shell at over 770 metres per second, showed itself in HWA tests able to penetrate 100mm of armour at 1,000m. In these trials the test plates were sloped at 30 degrees; against a flat target, it could go through 100mm at 1,500m. Armour-piercing shells used with the 88 contained a high-explosive charge designed to boost its destructive power once it had penetrated an enemy tank. It was discovered that its penetrative power could be further enhanced if the metal casing was thickened and the explosive reduced somewhat, the mass of steel being greater than the filling. Thus the 88 was made even more lethal, even if German scientists never followed this thinking to its logical conclusion, which would have been to go for solid armour-piercing shot (like that used by the British 17-pounder).

Firepower was thus the Tiger's outstanding characteristic. Its own armour – specified, coincidentally, to be 100mm on the front – would prove almost impossible for the T-34's 76.2mm armament to penetrate over 500m, thus setting the stage for so many Eastern Front duels where the German heavy was able to destroy the Soviet tank at two or three times the distance that it could return the compliment. Soviet gunners would try whenever possible to engage this terrible enemy from the sides or rear. But although the Tiger's lower flank armour was 60mm thick, its upper reaches (covering ammunition and fuel storage) was 80mm, which was also a very difficult challenge for Allied gunners at longer ranges.

Inevitably, given these levels of protection and the heavy armament it carried, there was a penalty to be paid in weight. Driven out of the factory, a Tiger was 52 tonnes. But 'bombed up' with a full load of fuel, ammo, tools and crew, it weighed in at 57 tonnes. For the design team, that meant a near doubling of weight in the space of two years' development. This brought a cascade of problems, from those relating to 'tactical mobility', i.e. its ability to move cross-country without sinking in soft ground, to 'strategic mobility', being the challenge of delivering it across long distances to the battlefield in the first place. Having appeased Hitler's desires for greater protection and firepower, those managing the panzer programme found he would wave them away when they raised the consequences for weight or size.

To stop this huge machine bogging down as soon as it went off-road, the army boffins at the HWA proposed a new solution. It was a system where the road wheels (those in contact, via the track, with the ground, bearing its weight) would be interleaved so that the tank's 50-plus tonnes would rest on 24 of them, spreading it more evenly. In time they concluded that even this was inadequate, increasing the number of road wheels to 32, which made the vehicle wider, in turn requiring the track width to be increased from 520mm to 725mm. Fitting broader tracks allowed Henschel and his designers in Kassel to reduce the Tiger's ground pressure from 1.545kg per square centimetre to 1.11kg. While improving the tank's cross-country performance, naturally, this came at a cost.

The production Tiger was 3.72m wide (compared to 2.62m for a Sherman or 2.88m for the Panzer IV), which meant that it wouldn't fit on railway flatcars or road transporters. In order to move it by rail, crews had to take the vehicle back one developmental stage, removing the outer road wheels as well as the mudguards, then fitting the narrower (520mm) tracks. Once they were unloaded close to the front, further backbreaking toil was needed to restore the vehicle to its combat configuration.

Having seen their original insistence on a 30-tonne weight limit blown apart, the HWA bureaucrats had in June 1941 laid one more onerous technical demand on the Henschel team: that the Tiger be able to plunge into rivers and indeed under water up to 4.5m deep. This 'fording' capability meant having to waterproof the vehicle and fit a snorkel device so that it could take in air and expel exhaust while underwater.

With the size and weight of the vehicle swollen to this degree, all manner of engine and transmission issues ensued. The 21-litre Maybach engine fitted to early models produced 650hp, giving a power-to-weight ratio of only 11hp to the tonne. The Sherman was a little better than this (around 13.5hp), but the T-34 boasted a figure of 18.9hp to the tonne.

Inevitably, given the multiple technical hurdles to its development as well as its complexity, the Tiger proved very expensive. That could be measured in financial terms, but also in the opportunity costs of making it. At 300,000 man-hours per vehicle, it took twice as much labour as a Panther.

There were 8,000 workers at the Henschel plant, the hulls being forged elsewhere before arriving in Kassel, where the suspension, engine and other components were added in one hall before the turret was fitted in another. If things were going well, each tank could be assembled in 14 days. It was quite a different process to T-34 or Sherman assembly, owing more to German traditions of precision heavy engineering than the Allies' application of automobile production techniques.

So we should not be surprised that the Tiger's price tag of 300,000 Reichsmarks compared to 103,000 for a Panzer IV or 117,000 for a Panther. At 1941 values and exchange rates, the US Army could buy three Shermans for the price of one Tiger. Arguably, though, the really significant price would prove to be the industrial consequences of following Hitler's obsession with 'super weapons' to its conclusion.

Historian Robert Forczyk has calculated that Porsche's commitment of its Alkett factory to the development and manufacture of Tiger (P) hulls during 1941–42 meant that it failed to produce something like 1,600 Panzer IVs it was contracted to make. Even allowing that 90 or so Tiger (P) hulls were converted into 'Ferdinand' heavy tank destroyers, the opportunity cost was enormous. And this calculation does not even factor in what might have been built, had the materials and manpower subsequently committed to Tiger production at Kassel been used to make less costly designs. In total, then, the development and production of 1,347 Tigers during the war was pursued at the expense of two to three times the number of Panzer IVs and Panthers.

There were other implications to the experimentation with more advanced weapons. As well as the multiple tank types in service, production began of turretless designs, for example three different vehicles mounting the long-barrelled 88mm gun. 'Hitler's decisions led to a multiplicity of parallel projects,' noted Albert Speer. 'They also led to more and more complicated problems of supply.'

Furthermore, General Guderian bemoaned 'constant modification to the design of tanks . . . countless variations to the original type . . . innumerable spare parts'. He could see the value in having some Tiger battalions for special missions but would prefer to have produced many more of the tried and tested medium panzers.

While he and other tank generals saw a logic for making Tigers, Hitler's belief that Germany should switch growing resources towards the production of various self-propelled gun models caused much grumbling. Guderian had to intervene in the autumn of 1943 to prevent a complete halt in Panzer IV production, as officials sought to use the chassis for more turretless tank-killers, something he believed showed 'a complete lack of comprehension of the real situation'.

Such was Hitler's personal interest in the Tiger project, and desire to micromanage its details, that it is pointless to dwell for too long on the counterfactual of what having so many more Panzer

IVs or Panthers might have done for the German war effort. Suffice to say, the price of making Tigers was considerable any way you measure it.

With a handful of Tigers delivered to the Eastern Front by September 1942, the Führer insisted they be thrown into action immediately. The attack, by a platoon of four tanks on the Leningrad front, quickly unravelled. As soon as they went off-road the heavy panzers sank into the soft ground, having to be destroyed by their own crews. Weary of his leader's interference, armaments minister Albert Speer recorded: 'Hitler silently passed over the debacle; he never referred to it again.'

Early in 1943, with numbers of delivered vehicles slowly building, the scene was set for action on a larger scale in Tunisia. In this campaign, British and American tankers got their first taste of how devastating the appearance of a Tiger could be. However, for the Panzertruppen of the two battalions using them, some downsides were soon apparent too. As an officer of the Grossdeutschland division, serving at that time in Russia, reported back to Berlin in spring 1943, 'the highly complicated Tiger must be as carefully maintained as combat aircraft in the Luftwaffe.'

When the battle was going the enemy's way and you needed to fall back, vehicles immobilized on mines or through breakdowns often had to be destroyed by their own crews, seven Tigers being lost this way during the Battle of Hunt's Gap in Tunisia in March 1943, for example. They were so heavy that there were very few Wehrmacht vehicles capable of towing broken-down Tigers away. On the Eastern Front, driving them long distances produced frequent breakdowns and long hours of maintenance.

Technical issues, slow initial deliveries (it took until May 1943 for Tiger production to reach 50 per month) and two-front warfare (sending dozens to Tunisia) meant that one return for April 1943 showed only 31 of the tanks operational on the entire Eastern Front. Thus, during its early months of service, the period of its greatest technological superiority over Soviet designs, there were precious few Tigers where they were most needed.

By the time of the Battle of Kursk in July 1943 production had increased, so the Germans were fielding battalions with twice as many Tigers as those used in Tunisia, in the hope that a significant number of working vehicles could be kept going. During Operation Citadel, the opening Kursk offensive, the 505th Heavy Tank Battalion would claim 110 kills, but it did so with an availability rate that averaged just 45.7 per cent during two weeks of fighting, with the number of working Tigers at times falling as low as six.

Germany's enemies drew their own conclusions about this super weapon. Tank crews in the field were often terrified by it. But from staff men a little further from the front, there were more sanguine judgements. Writing after the British captured their first working example, the famous Tiger 131 (now at The Tank Museum, Bovington), a British intelligence officer noted: 'The tank bristles with every sort of complication, and one would think that it would be at least twice as difficult to produce as either of its predecessors. This may have a bearing on the numbers that are likely to be met with in the future.'

As for the difficulties of stopping one of these heavy tanks, the capture of Tiger 131 in Tunisia had shown that existing weapons might be used to good effect. In that instance a platoon of Tigers was moving to attack British infantry, the 2nd Battalion the Sherwood Foresters, on a hilltop. As we saw in the last chapter, they had fired a brand-new weapon at it – the Projector, Infantry, Anti-Tank or PIAT – but to no effect.

Shoulder-launched arms like the PIAT, hurling bombs that used shaped charges that were able to penetrate even thick armour, were a significant part of the latter war story, giving the ordinary foot soldier the chance to knock out tanks. On this day in April 1943, however, they did not provide the answer to the seemingly unstoppable progress of Tiger 131. Instead, it was a Churchill tank supporting the Foresters that did it with a 6-pounder shot at a weak spot, just where the turret met the hull, jamming the Tiger's traversing mechanism. Unable to pan the gun in order to knock out the Churchill, the German crew bailed out and surrendered, leaving their intact vehicle to be captured.

Even soon after its debut, then, the Tiger was not invulnerable, but it did enjoy a big margin of superiority over Allied tanks. Some 17-pounder anti-tank guns were already in the field and could deal with it, but as far as the crews of Shermans were concerned, there was consternation at the advice they were given by their commanders. Exploiting the fact that the 75mm high-explosive shells fired by their guns could be fused to explode after a fraction of a second's delay, they were told to aim a couple of hundred yards in front of the Tiger in order to ricochet the shell off the ground. The idea was that when it exploded it might do so over the panzer's top, sending shrapnel through any open hatches or damaging the crew's vision devices. These techniques were practised by Sherman crews (and had some use against other targets, such as infantry in trenches) but they knew only too well that the odds of taking out a heavy tank by this method were very long indeed.

Those Shermans later armed with the American 76mm gun had a slightly better chance of taking on a Tiger. Dmitri Loza, commander of a Soviet army M4A2, described an ingenious tactic. The challenge was how to get the Tiger to slew around to expose its weaker armour: 'The Sherman could never defeat a Tiger with a frontal shot. We had to force the Tiger to expose its flank. If we were defending and the Germans were attacking, we had a special tactic. Two Shermans were designated for each Tiger. The first Sherman fired at the track and broke it. For a brief space of time the heavy vehicle still moved forward on one track, which caused it to turn. At this moment the second Sherman shot it in the side, trying to hit the fuel cell.'

These methods, whether trying to skip a shell over it or expose the flank, speak to the disadvantages that Allied tankers tried to surmount, and the quality of the Tiger's design. No wonder a good many of them pulled back or even bailed out when faced with this awful challenge.

Given these factors, naturally the Allies were doing all they could to further complicate production by bombing German factories. In October 1943, the Royal Air Force dropped 1,800 tons of bombs on and around the Henschel plant, disrupting output for a few weeks. Tiger production would peak in April and May 1944 when around 100 were made in each month. Of course, this was a very modest figure compared to the Panzer IV, which peaked at more than three times this late in 1943, let alone the enormous Sherman figures.

That the numbers were scarce, and there was a need to deploy heavy tanks very carefully, was all too evident to Wehrmacht leaders. A directive on the use of Tigers noted that they were only to be used 'in the decisive point of the battle in order to force a decision . . . they are especially suited for fighting against heavy enemy tank forces and must seek this battle.'

In June 1944, the British army would receive an awful lesson in what could be achieved when such orders were followed to their full effect.

As far as the English-speaking world is concerned, the greatest demonstration of the Tiger's power came in the rolling Normandy countryside on the morning of the 13th of June 1944. One week after D-Day and the opening of the long-awaited Second Front against Hitler, British generals saw a gap opening up in the German defences, throwing the 7th Armoured Division into it.

The Desert Rat veterans initially advanced quickly, and by mid-morning had paused their progress along a main road (the N175) that ran through the town of Villers-Bocage. Lead elements from A Squadron, the 4th County of London Yeomanry (CLY) had gone beyond it, towards the key city of Caen, pausing atop a ridge that dominated the surrounding countryside.

Just over half a mile behind and below was the rest of their regiment, lining the N175 as it passed through Villers-Bocage itself. Next, on the same road as it descended a ridge into the town, was the second regiment in the division's order of march, the 5th Royal Tanks (RTR).

What none of these British soldiers knew, as they paused their advance, many of them jumping out of their vehicles to get a brew of tea on, was that parked in a hide area just off the N175 as it mounted the ridge beyond Villers-Bocage was a Tiger company under SS Hauptsturmführer (Captain) Michael Wittmann. They had heard and glimpsed A Squadron of the 4th CLY sailing past. In his after-action report Wittmann noted: 'I had no time to assemble my company; instead I had to act quickly as I had to assume that the enemy had already spotted me and would destroy me.'

Leading four panzers forward, Wittmann launched a devastating attack. Approaching the N175, the A Squadron group was to their right, and the remainder of the 7th Armoured Division on the left. The first thunderclap of an 88 firing at close range was aimed at an A Squadron tank. One by one the Germans brewed up their Cromwells and Sherman Fireflies, preventing the lead squadron from turning back to take part in what was to come. It was then that Wittmann, accompanied by a second Tiger, turned and started to rumble downhill towards the town itself.

'I drove toward the rear half of the column . . . knocking out every tank that came towards me as I went,' he wrote. 'The enemy got thrown into total confusion.' Many of the British crews, it will be remembered, were not even in their vehicles, some men scattering into nearby fields or streets, as the Tigers went to work.

At the entrance of Villers-Bocage, however, a couple of Cromwell crews, having heard the developing battle, were waiting to meet Wittmann. 'Major Carr, the 2 i/c, fired at it with his 75mm,' recorded one CLY tank commander of one of the engagements at no more than 50 metres, 'but heart breaking and frightening, the shots failed to penetrate the [Tiger's] side armour even at this ridiculous range.' Panning his 88 onto the target, Wittmann's Tiger fired, quickly dispatching the threat.

There were quite a few British weapons in that Norman town that could have stopped the slaughter, from Firefly tanks to 6-pounder anti-tank guns and PIATs. But for an hour or so Wittmann was able to run riot, in large part due to the unprepared state of the 4th CLY. Discussing it decades later with Roy Dixon, adjutant of 5th RTR on that day, who watched the whole disaster unfold, his cold fury with the Yeomanry's commanding officer was still evident when he told me: 'he should have been court-martialled.'

Wittmann's attack was eventually stopped when one of the British infantry gun crews, waiting for his Tiger to pass, slammed a 6-pounder shot into its rear, immobilizing the vehicle. The SS officer escaped by foot. His assault had cost the 4th CLY 25 tanks (18 Cromwells, four Fireflies and three Honey lights). Their accompanying infantry from the 1st Battalion, the Rifle Brigade had lost 14 halftracks, eight Lloyd carriers and eight Universal carriers.

That afternoon, another Tiger company was sent into Villers-Bocage and attempted to exploit to its south. The psychological effect of the morning's battle had struck deep. On sighting Tigers moving towards them, three Cromwell crews from 5th Royal Tanks had abandoned their vehicles. The Tiger's appearance, its sheer size as well as the 88's menacing muzzle, all contributed to the psychological effect.

Sergeant Jake Wardrop of 5th RTR, a Scottish veteran of the Desert Rats' North African battles, was commanding a Firefly that day. Glimpsing an approaching Tiger, he wrote later in his journal that it was 'as big as two Glasgow corporation tramcars side by side'. His tank opened up with its 17-pounder, but the gunner was too scared to aim properly. However, in close-range fighting many panzers were disabled in Villers-Bocage, and heavy bombing by the RAF flattened the town, preventing the recovery of some immobilized vehicles like Wittmann's own.

During the afternoon of the 13th of June and the subsequent day, German records show, nine Tigers in Wittmann's battalion were destroyed and six seriously damaged (i.e. most of those engaged on the 13th/14th of June). Additionally, around ten Panzer IVs of the supporting Panzer Lehr Division were destroyed and hundreds of their troops became casualties. So, the outcome of 30 British tanks taken out was quite different to some of the crazily unequal Eastern Front tallies, not least Wittmann's destruction of dozens of Soviet tanks at Prokhorovka for the loss of a single Tiger from his platoon. American historian Christopher Wilbeck has noted: 'Of the nine Tigers that entered the town of Villers-Bocage throughout the day's battle, including Wittmann's Tiger, all but one was either damaged or destroyed.'

All of this, though, risks losing sight of the bigger picture. Even if one believes that the 101st SS Heavy Tank Battalion was essentially sacrificed in this battle, the effect that its fewer than 20 working Tigers had on the wider campaign was dramatic. A bold attempt to break out of the Normandy beachhead using a veteran armoured division had been thwarted. Although the 7th Armoured Division smashed a series of German counter-attacks on the 14th of June, it had to withdraw that night, losing the advance it had briefly gained. Thereafter the Allies had to slug it out for several more weeks before they could do this.

The fate of the 4th CLY and indeed Wittmann's battalion, while demonstrating the Tiger's power, also speaks to the human factor in war. As we've seen, the leading British elements had quite a few weapons that *might* have blunted the initial attack, had their crews not been taken by surprise and their leadership not failed. That the attack was the success it was attested not just to British shortcomings but to Wittmann's aggression and confidence. And indeed it was that last factor, arrogance some called it, that led to a very different outcome when his Tigers met Allied tanks several weeks later on the 8th of August.

By that point in the Normandy campaign, matters were going very badly for the Germans, and the Allies were breaking out. Patton was leading his Third US Army to the south and west of the defenders as General Eisenhower sought to encircle them. Meanwhile, to their north, Montgomery launched Operation Totalize, a thrust by British, Canadian and Polish divisions designed to pin their enemy frontally while helping to cut their line of withdrawal to the east.

The battle that took place south of Caen on the 8th of August was a local counter-attack by SS battle groups, trying to prevent this. The 101st Heavy Tank Battalion could by this point muster eight to ten Tigers, and Wittmann was sent from the village of Cintheaux towards Saint-Aignan, a couple of miles to the north, which had just fallen to a British attack.

Wittmann chose to advance up a depression in the ground stretching northwards, with Saint-Aignan to his front right, and another village, Gaumesnil, ahead and to the left, where Canadian troops were establishing positions. Seeing the Tigers advancing, a Canadian officer watched them coming through the fields, feeling 'astonished at the cool arrogance of the German tank commanders, standing up exposed in their turrets looking for targets through their binoculars, their guns traversing all the time'.

To the south, another spectator, Kurt 'Panzer' Meyer, the SS divisional commander who had dispatched Wittmann, observed from the outskirts of Cintheaux. 'The Tiger of Michael Wittmann races into the enemy fire,' Meyer later wrote. 'I know his tactics during such situations, it is called: Straight ahead! Never stop! Get through and gain an open field of fire!'

In an orchard just south of Saint-Aignan, the crew of a Firefly from 3 Troop, A Squadron, 1st Northamptonshire Yeomanry got the word to engage the targets advancing across their front. In the gunner's seat was Trooper Joe Ekins, a shoemaker by trade. His description of what happened next gives an excellent sense of the tactics they adopted: 'Someone saw these three tanks which were coming across our front and I was the only tank that had got a gun big enough to do it . . . They were coming across at 1,200 yards. The officer decided that we'd wait until 800. So we pulled forward into a firing position and in fact I could see all three of the tanks now coming across . . . [Sergeant Gordon] said "line up on the back one", the rear of the three which again was standard sort of practice, once you've got the one at the back then you've got a chance to get the other two. And we did and at 800 yards I got the command to fire, and I fired two shots and knocked him out.'

Double tap, scratch one Tiger.

The Firefly commander, Sergeant Gordon, ordered them to reverse out of the firing position so they could reload somewhere the enemy couldn't see them. Wittmann called out on the radio, 'Pak right!', assuming the fire had come from a Pak or anti-tank gun. One of the Tigers panned its gun right, firing at Ekins's tank as it reversed into the trees.

The 88mm shell passed just over the Firefly's turret, possibly wounding Gordon. Ekins's account implies that his commander's nerve went at this moment: 'whatever happened, he jumped out anyway and I never heard from or saw him again, but the troop officer jumped in and took command of the tank.' This young officer, Lieutenant James, ordered the Firefly crew into a new firing position (in case the enemy was still aiming at the first one), and told Ekins to line up on a second target.

Moments later, Ekins related that he 'fired at the second one, one shot, and knocked him out'. Two Tigers down. Again, the Firefly backed out of its firing position on the orchard's edge – British crews call this 'jockeying back' – before driving to another vantage point: 'we . . . came out again and fired two shots at him.' Ekins had struck again: three Tigers destroyed in 12 minutes. His acting squadron commander declared it had been 'like a day on the range . . . rather fun shooting'.

This was frivolous, for the Northants Yeomanry would lose 20 tanks of their own that day. The main arm of Meyer's counter-attack, with 30 armoured fighting vehicles (mostly Panzer IVs) and 200 *Panzergrenadiers*, then hit the southern part of Saint-Aignan. During this fighting one of the Panzer IVs, using folds in the ground more carefully than Wittmann's company, knocked out the Firefly that had dealt with the Tiger column. Nevertheless, the British and Canadian troops held on to their respective positions.

During the initial engagement, a total of five SS Tigers had been destroyed. Wittmann's vehicle suffered a catastrophic hit; internal explosions blew off the turret and claimed the lives of the whole crew. A Canadian armoured regiment in Gaumesnil had taken part in the action, as had another British one. So was Wittmann's tank one of the three destroyed by Ekins or was it hit by one of the others? The issue remains a lively historical controversy.

It certainly wasn't the first Tiger that Ekins hit. A survivor from, that crew, Hauptscharführer Hans Höflinger, later reported that while seeking cover after bailing out, he had seen 'the wake of a shell going through the barley, and it hit Hauptsturmführer Wittmann's tank. It came from the right.' This could imply that Ekins, who was off to their right, killed Wittmann's crew during his second or third engagement, but in truth we cannot be sure.

For our purposes, it is better to draw conclusions about how far superior tactics and crew discipline can compensate for, or even outweigh, the technological factors. At Villers-Bocage, Wittmann's arrogance served him well. At Saint-Aignan it cost him and others their lives.

Moreover, the Normandy battlefield was very different to that of Prokhorovka one year earlier. The built-up nature of the French countryside, its many hedges, farms and undulations, meant engagements took place at ranges that made it harder to exploit the Tiger's hefty punch. And of course, by 1944 the British had deployed the Firefly and other anti-tank weapons that could deal with Tigers, even when they were more than 500 metres away.

In the battlefield circumstances of the 8th of August at Saint-Aignan, and under more cautious commanders, a Tiger, Sherman Firefly or Panzer IV were all on a level, in the sense that each could destroy the other, even 800m or perhaps 1,000m away. And of course it was a Panzer IV that knocked out Ekins's Firefly. That idea takes us back to the question of whether the Tiger's high cost made it the right choice.

Would manufacturing two or three Panthers or Panzer IVs instead have been a better use of resources? And given the difficulties of maintaining the Tiger compared to those other two tanks, might making more medium panzers actually have translated into an even higher number, maybe three or four, actually in the line of battle? As the war neared its end, the failure of Germany to match American or Soviet production levels became more and more apparent.

Operation Totalize, coordinated with the American sweep to the south, marked the breakout from the Normandy beachhead. German divisions trying to escape the Allied encirclement streamed eastwards, while Typhoons and Thunderbolts dived down on them, torching one column after another. Some of the Tiger crews got as far as the River Seine but, unable to find a means of crossing, disabled their vehicles there.

When British staff officers picked their way through the carnage of the Falaise pocket, they found 27 Tigers lost by their enemy during the 8th–21st of August. Just one of these was reckoned to have been knocked out by armour-piercing shot; 20 had been destroyed by their own crews, and six abandoned before they could do so.

Among the wrecks were several Tiger Bs – dubbed King or Royal Tigers by Allied troops. This was a new design that Henschel started building in Kassel early in 1944, allowing a couple of companies to be fielded in Normandy. By the late summer, production of the original Tiger had stopped, with 489 King Tigers being made before the war's end.

This new vehicle took the Führer's obsession with technical superiority to its logical, absurd conclusion. Its long 88 was even more powerful, its armour still thicker, giving a combat weight close to 70 tonnes. Improvements in the Maybach powerplant were insufficient to offset this, leading to greater mobility issues even than those suffered by its predecessor. At 800,000 Reichsmarks, the Germans could have made seven Panthers for a similar price.

With Allied forces advancing into Germany, Henschel kept producing vehicles until very near the end, handing 13 tanks to the Wehrmacht in April 1945, just three days before Kassel fell. At this moment of chaos, the dying days of Nazism, Erwin Aders, their chief designer, chose to write an account of the Tiger's development.

His apologia for the vehicle's shortcomings slyly put the onus on subcontractors and the exigencies of war: 'Failures and breakdowns at the front – which frequently led to voluntarily blowing up brand-new vehicles – should be considered to be due almost exclusively to deficiencies in procured components. The rest is to be blamed on the fact that mass production had to start immediately, before test results were available; a point that can't be stressed often and strongly enough.'

So, just as the Tiger in many respects foreshadowed the Cold War philosophy of trying to offset Soviet mass with Western quality, Aders showed the way in terms of assigning his company's failures to the customer and suppliers. Of course, from the design and operating point of view there was much to be proud of, and he trumpeted the Tiger's success also. Even in the last months of the war, the appearance of one of these panzers could hold up an Allied advance for hours.

In March 1945, General Eisenhower solicited the opinions of American tank crews about their experiences. He knew that reports of the Sherman's inadequacies had prompted politicians to ask questions back home. In Britain too, there had been a 'tank debate' in Parliament the previous summer.

'It is my opinion that the [Panther] and [Tiger] enemy tank is far superior in manoeuvrability to our own Sherman tanks,' Captain Henry Johnson reported to Ike. Staff Sergeant Alvin Olson picked up on the same point, noting that, even though Tigers were much heavier than the Sherman, their wider tracks and the design of their running gear meant they could move freely in places where 'the tracks of our own M4 tanks were often deep enough in the same field to show marks of the tank's belly dragging.'

As for armoured protection, another Sherman commander, Sergeant Harold Fulton, described blasting a column of six Tigers and two Panzer IVs that was 500–800 yards away: 'Along with my gun firing, there were four more tanks of my platoon. Two or three M4 tanks from another company and two M7s firing at the same column . . . but of all of the Mark VIs [i.e. Tigers] there was one penetration in one tank.'

Eisenhower digested these reports, concluding it inevitable that in the compromises needed to field a tank with the mobility that his way of war required, tankers would inevitably feel hard done by. 'When a man is actually on the front line engaging a gun or a tank, he could not have, in his own estimation, a big enough gun or enough armor,' Eisenhower wrote to his boss, General George Marshall. Following the tank soldier's desires to their conclusion, Ike felt, would produce 'a steel pill box'.

Marshall himself produced probably the most eloquent defence of the very different philosophies that governed American and German tank design. He freely admitted that 'the German Tiger and Panther tanks outmatched our Sherman tanks in direct combat.' But Marshall's case for the Sherman emphasized strategic mobility – the need to ship it across the world, then move long distances over land – and America's very different doctrine: 'Our tanks could not well be of the heavy type. We designed our armor as a weapon of exploitation. In other words, we desired to use our tanks in long-range thrusts deep into the enemy's rear where they could chew up his supply installations and communications. This required great endurance, low consumption of gasoline and ability to move great distances without breakdown.'

Would the Germans have accepted this? Evidently that depended on who you asked. General Hasso von Manteuffel, commander of a number of panzer divisions and armies during the war, summed up his views thus: 'Tanks must be fast, that I would say is the most important lesson of the war in regard to tank design . . . we used to call the Tiger a "furniture van" – though it was a good machine in the initial breakthrough. Its slowness was a worse handicap in Russia than in France because the distances were greater.'

Perhaps there was a sense in which tank soldiers in all armies emerged from the war envying their enemy's kit. When the

Germans came to design their Cold War tank, the Leopard, as per Manteuffel, they placed a distinct emphasis on its mobility at the expense of armoured protection. But that British product of the 1960s, the Chieftain, did the opposite.

Arguably, the experience of fighting the Tiger left Allied soldiers with the view that firepower was the prime characteristic of the tank, leaving open the argument about how the trade-offs between protection and mobility should be made.

Certainly, it was evident that no Second World War tank was truly invulnerable. There were some ways of knocking out a T-34 in June 1941, or a Tiger early in 1943, even if their appearance struck fear into the enemy, forcing the rapid development of more effective countermeasures.

Even in 1944 or 1945, then, the task facing those designing the tanks of the future would be to find a level of protection that could guard against the majority of threats the vehicle might encounter on the battlefield, and a gun that could destroy most or all of the machines that might be ranged against it.

Victory against the Nazis was marked by celebrations and many a fraternal toast. But among the Western officers attending the Red Army's parades, eyes soon fixed on the new JS-III or Stalin heavy tank. No sooner had the guns fallen silent than a new rivalry blossomed, and with it the rumbling onward of military technology. The rules would be different, but in the Cold War the resources committed to producing the perfect tank would grow ever larger.

Centurion

Entered service: 1946
Number produced: 4,442
Weight: 50 tons (51 tonnes)
Crew: 4
Main armament: 20-pounder (84mm) gun, later 105mm gun
Cost: £35,000 (£1.15m at 2024 prices); however, later figures of up to £60,000 per tank are also cited

Centurion

Dawn on the 25th of April 1951, and the scene was set for one of the darkest days in the British army's history. A single brigade, the 29th, holding a group of hills on the banks of Korea's Imjin River, north of Seoul, had been assaulted by an entire Chinese army, the 63rd, and it was breaking. Their Communist enemies had come forward during the night, leaving many of the British troops surrounded. The enemy offensive had started three days earlier, and now disaster stared them in the face.

Private William Gibson of the Royal Ulster Rifles was one of those ordered to withdraw in the dawn mist. They expected to fall through the Royal Northumberland Fusiliers, as the brigade concertinaed back in an orderly withdrawal. Reaching their ground, instead 'we discovered it was deserted . . . the word was that they'd already pulled out during the night.'

With thousands of Chinese troops infiltrating their positions, and in an attempt to stabilize the situation, the brigade commander had sent forward ten Centurion tanks of C Squadron, the King's Royal Irish Hussars. They were tasked with covering the infantry's withdrawal from two parallel valleys.

On the 24th, tanks from C Squadron had tried and failed to get through to the 1st Battalion of the Gloucestershire Regiment, via the western of those two features. 'A half squadron had tried to get up to them and couldn't get up,' Corporal Dennis Whybro would later recall, noting that the enemy had infiltrated beyond the Centurions even while they made the attempt: 'you had to fight your way out to come back and bomb up, they were behind you as well as ahead of you . . . there were so many Chinese that you couldn't kill enough of them.'

Now again on the morning of the 25th, the Centurions were in action, the valleys echoing to the blast of their shells and constant chatter of machine guns. Along the valley floor, a road designated by the military as a Main Supply Route or MSR became the focus of the battle. Private Gibson, walking down towards it, saw 'Centurion tanks on that MSR firing forwards towards the Imjin river crossing . . . they were firing at a great rate.' Turning around to look behind him, the infantryman was horrified to see 'a mass of figures running forward, in waves running forward, Chinese!' The tank crews tried to hold them back, Corporal Whybro saying: 'you could see the Chinese walking along. Endless chains of them. It [didn't] matter what you done, they seemed to keep coming.'

At 9.30 a.m., the commanding officer of the Glosters, surrounded and without any hope of relief, had given the order for his soldiers to attempt a breakout. Not long afterwards, a few miles to the south-east, parties from the Royal Ulster Rifles, wrapped up in their own drama, heard 'the order [come] through, "every man for himself, get out as best you can", head south!'

British officers were trying to use all manner of vehicles to assist their escape. There were Universal and Oxford carriers (tracked machines used for infantry support), halftracks and the Centurions. Soldiers clambered onto all of them in an attempt to outrun their pursuers. But the Chinese were stretching across the cold, barren Korean hillsides as well as along the valleys, reaching positions where they could fire down on the broken brigade below. Edward Tyas, another member of the Royal Ulster Rifles, followed those racing towards the tanks, hoping for a ride out, but then saw 'the Chinese were picking them off like clay pigeons . . . there was carnage on the valley floor.'

Scores of stragglers were being taken prisoner. As the day wore on, the Chinese were poised to capture tanks too, a matter of great concern to the British, since the Centurion was the most modern tank in the Western armoury. Indeed, soon after their arrival in Korea in November 1950, the War Office had ordered the emergency withdrawal of this brand-new armour, fearing that these hitherto secret weapons were about to fall into Communist hands. That order had been rescinded, but on that April day a few months later, advancing Chinese troops were trying to board the Centurions.

Private Gibson of the Ulster Rifles, still hoping for salvation, watched 'groups of Chinese running up, jumping up and putting explosives on the tanks, suicide squads'. Spotting a place where the MSR kinked in an 'S' bend so traffic had to slow down, the Chinese set up an ambush position, using small arms, 'sticky bombs' and pole charges (demolition explosives that could be thrust at an enemy vehicle) to disable a halftrack and an Oxford carrier.

Moving into the ambush zone, one of C Squadron's Centurions also fell into the trap and was immobilized. It was here that the fighting assumed its most desperate character. Leading a group of Centurions forward, Captain Peter Ormrod of the Hussars directed them off-road in order to mount a counter-attack across the rice paddies.

The enemy was so close that commanders who had been taking potshots or hurling grenades at the Chinese closed down. With hatches shut, crews relied on vision blocks or periscopes to conduct these manoeuvres. This posed its own challenges, and Ormrod's tank soon got stuck fast at the edge of a paddy.

Sensing their moment, Chinese infantry rushed towards the Centurion, a couple managing to board it before one of the Hussars' following tanks 'hosed it down' with machine gun fire, cutting down the attackers. Thus saved, Ormrod's driver managed to dislodge his 'Cent' and reverse away from the bank. However, another tank had become immobilized through shedding a track. Where they had the chance, the tank crews blew up their vehicles once they'd bailed out. It had been impressed upon them that they could not allow the Chinese to take a working Centurion.

South of the S bend, the scene on the MSR was by early afternoon one of chaos. Burning or driverless vehicles, including ambulances with casualties aboard, had been abandoned. Parties of British and Chinese troops were wandering about, exchanging fire as others made their final bid to escape. The last couple of tanks to withdraw had to smash their way past other, disabled vehicles. Major Henry Huth, C Squadron's officer commanding, called it 'one bloody long ambush'.

Private Gibson watched a Centurion, loaded with soldiers, approaching a spot where the Chinese were waiting on both sides of the road: 'I could see the burp gunners [sub-machine gun armed troops] just firing on top of the tank and it was covered with people hanging on, literally they were raking them from side to side on both sides. The heavy machine gunner was firing from the other side. You could actually see the people on top bouncing as the bullets hit them.'

The tanks had traversed their guns rear, towards most of the enemy, as they tried to get out. But there were so many friendly troops riding on the back decks that they didn't dare keep firing in case they hurt them. Chinese rounds, though, were taking such a toll of passengers that one Centurion reportedly had blood running down the bazooka plates on its sides. Ernest Pilbeam, another King's Royal Hussars crewman, noted that 30 infantry had ridden out on one tank, but 'by the time the tank got back to the bivvy, back to the other tanks, there were only about 5 or 6 men left on there alive.'

So ended the Battle of the Imjin River, which cost the British more than 1,300 casualties, 141 killed. The 'Glorious Glosters'' last stand on Hill 235 was lionized in the press back home, 522 members of the battalion becoming prisoners of war. That fate befell William Gibson from the Ulster Rifles too, left behind when he realized that he couldn't hitch a ride out. As for the Chinese casualties, they were estimated as many thousands.

The 29th Brigade's delaying action fought over three days had at least helped to check a headlong rush to the south following China's entry into the war. Thereafter, the tanks deployed in support of United Nations forces were kept clear of the '*sauve qui peut*' dramas of the 25th of April, assuming more often the role of artillery support in what was largely an infantry battle.

The laurels from that dreadful battle included two Victoria

Crosses and one George Cross to soldiers from the Glosters. In the 8th Hussars, the squadron commander was awarded the Distinguished Service Order and Captain Ormrod the Military Cross. Undoubtedly the tanks had helped hundreds to escape captivity by suppressing the Chinese so effectively. Five of the ten Centurions sent forward were left behind – although when Allied troops recovered the valley they were relieved to find that none of the disabled vehicles had been removed by the Chinese.

The *Newcastle Journal*, reporting the battle one week later, headlined its article: 'RNFs Saved by the Hussars Tanks'. They had used the fate of their local regiment, the Royal Northumberland Fusiliers, as a hook for readers, but the article went on, 'it is clear that one regiment and one regiment alone saved the Northumberland Fusiliers, the Royal Ulster Rifles, and the Belgian battalion from sharing the Glosters' fate. That regiment is the 8th Hussars.' If any general or Whitehall bean-counter had been wondering about the continued utility of tanks in the nuclear age, the soldiers' testimony underlined their unbridled gratitude.

As for that new machine, the Centurion, it had made a convincing debut, becoming beloved by its crews. They had enjoyed that comforting sensation of knowing that the great majority of the enemy's weapons could not harm them. Corporal Whybro's verdict was that 'you couldn't fault the gun, you couldn't fault the engine.' Another 8th Hussar who came through those fights remembered: 'being surrounded by that armour I felt completely safe, even on the two occasions we were hit.'

There was just a single engagement between a Centurion and another tank during the entire Korean campaign. It happened a couple of months before the Imjin battle when a Cromwell captured from the 8th Hussars was turned against them by the North Koreans. It was swiftly dispatched with two shots from a Cent's 20-pounder.

As for its other qualities, the Centurion was slower than a Sherman on-road but demonstrated superior mobility off it, being less likely to bog, and climbing relentlessly up the many hillsides. 'We had the most advanced tank in the world at that time, the Centurion,' Trooper William Bye of the Hussars believed. 'As a result we could take a tank where tanks wouldn't go.'

While development had started during the dark days of the war, with many seeing the design as a British answer to the Panther, it was that bit more modern than the Sherman and Pershing models being used by the Americans. However, the truly remarkable thing about the Centurion was not that it had an edge at the start of its career but that it had such longevity. It was to prove so useful to armies around the world that it played a major role in Cold War conflicts between 1950 and 1991. Indeed, even in that final year, just months before the collapse of Soviet Communism, I watched Royal Engineer variants of the tank in action during the liberation of Kuwait. Its story, then, is not of a revolutionary vehicle but of one so well-conceived and adaptable that it was able to survive the evolution of technology for such a long period. This qualifies it as the greatest of British tanks – and yet it emerged during the period of muddle and dysfunction when many soldiers had lost all faith in the country's ability to design such a machine.

Those in the British army who despaired at its wartime inability to make decent armour often date the arrival in 1942 of its Department of Tank Design at buildings in Chobham Lane, Chertsey as the moment a turnaround began. There, on a site about 30 miles south-west of central London, they shared digs with the tank boffins and those who procured fighting vehicles.

Step by step, the different interests in the making of war machines were herded together in a Tank Board, the chair of which, a naval officer, Commander Robert Micklem, was also in charge of the department within the Ministry of Supply which tasked the manufacturers. Thus, the engineers, designers, procurement types and pen-pushers were strapped in harness. The fighting soldiers were also represented by a brigadier made commandant at Chertsey and the major general running the procurement of armoured vehicles.

During the first few years of the war these many different players had struggled to untangle the web of vested interests – military, engineering and industrial – that led to the production of some dreadful tanks. Peter Beale, a veteran of the 9th Royal Tank Regiment, was so angry at the consequences that he subsequently tried to explain how it happened in a book titled *Death by Design*. On a similar note, The Tank Museum's curator, David Fletcher, called his *The Great Tank Scandal*.

Army politics, the fights between modernizers and traditionalists, Tank Corps and cavalry men, had in the 1930s produced the doctrinal division between 'infantry' and 'cruiser' tanks. The former, intended to be heavily protected, would assist the infantry to break through in the best traditions of Cambrai. The latter were meant to exploit that victory, dashing forward into the enemy rear, and therefore were more lightly armoured and faster.

So, the Matilda or Churchill represented the infantry tank concept, while the A13, the Crusader or latterly the Cromwell were cruisers. Other armies had their own doctrinal kinks, but they really couldn't see the point of this type of divide. Thus, in the US Army, infantry divisional tank battalions and the 'exploitation' forces, the armoured divisions, used exactly the same machine, the Sherman. Similarly, there were the Soviet T-34 or German Panzer IV, although both of those armies also had specialized, turretless assault guns to bolster the infantry.

As the Sherman proliferated in British service, it played both roles and quite a few people, including General Bernard Montgomery, started wondering why there was any point in maintaining distinctions between tank types. So, the yearning for a 'universal' or 'capital' tank design was heard, the Tank Board opining in September 1942: 'the British and American staffs are in agreement that the major requirement is for an "all-purpose" tank, the standard components of which should provide the degree of flexibility required to mount the various type of tank armament in use or under development.' One of the advocates of this US/UK project even dubbed it the 'Victory Tank'.

This, however, touched on deeper sensitivities, passions beyond the ken of the slide-rule-carriers or pen-pushers, for some of the cavalry generals, seeing that mechanization had finally killed off their beloved horses, looked to the 'cruiser tank' role as the justification for their survival once the war was over. For other officers there was simply a fear that if infantry and cruiser roles merged, both roles would suffer. So, one year after the aspiration for an 'all-purpose tank', the General Staff restated the orthodoxy, defining the need for a new 'cruiser', designating the vehicle the A41. And this was the vehicle that became the Centurion, a journey that would evolve into precisely the type of all-purpose or universal tank that some of the more political generals didn't want but that engineers and fighting soldiers knew was essential.

Credit for diverting the project from yet another under-armoured, under-gunned British cruiser into something altogether better belonged largely to Claude Gibb, an Australian engineer who had been co-opted into the Ministry of Supply in 1940. He conformed to the Poms' stereotype of an Aussie by being a straight-talker. Once Gibb was put in charge of armoured vehicle procurement, all of the earlier doctrinal thinking was left behind. 'I think of him as a tornado of fresh invigorating air coming into an organization hamstrung by tradition and lack of organization,' said one engineer who worked with Gibb, adding that their Aussie boss did not chide or coerce but rather relied upon 'outstanding personality, drive, unbounded optimism, and cheerfulness'.

Between the General Staff's specification for a new cruiser issued in September 1943, and the Tank Board's revision of it just a matter of a few weeks later, many of the key attributes of the A41 were changed: the target weight went up from 40 tons to 45 (and would end up at 50); the preferred armament was changed from the 75mm all-purpose gun to the 17-pounder (but with scope to fit heavier weapons too); the width restriction required for British railway flatcars was waived; and the practice of mounting a machine gun in the hull was abandoned so that the vehicle's frontal armour could be sloped for better protection.

The experts at Chertsey had examined the captured Tiger 131 in detail, and by autumn 1943 and the refinement of the A41 design, they were also getting information from the Red Army about the Panther, following its debut at Kursk. Whether it was the latter's sloping frontal armour, or the widening of a tank in order to mount a larger turret and therefore a bigger gun, they learned lessons from German design.

While some characterized the Centurion as the 'British Panther', an idea which had some justice as far as the arrangement of armour was concerned, it was also meant to be able to slug it out with a Tiger. Thus, in 1944 it was specified that the frontal armour should match the Tiger's 100mm (though the Centurion turret's armour was quite a bit thicker than that) and that its gun should be able to defeat the German heavy. And not long after the vehicle entered service, it was up-gunned with a 20-pounder (or 84mm gun) intended to deliver a similar punch to the Tiger's 88.

Like so many of the vehicles we've looked at, armour thickness crept up during development, hence weight increased and improvements were required to maintain mobility, such as fitting wider tracks as well as boosting engine power. It took just under 18 months from the specification of the vehicle to the shipment of the first prototypes across the Channel in May 1945, just after the fall of Nazi Germany.

As the weight climbed, compromises had been made. As with the Panther, the designers had to accept that protection on the flanks and rear would be thinner. And without further increasing the internal volume, with all that implied for size and weight, they'd had to settle for smaller fuel tanks than would have been ideal.

The Meteor engine, chosen because it met the specification requirements for reliability, 'simplicity of operation and ease of maintenance', was a reassuringly mature design by this point, but it was very thirsty. The original 1945 trials in Germany suggested that the tank delivered 0.7 miles to the gallon. However, users later gave lower suggested figures. On-road, it could burn through its 120 gallons (550 litres) of petrol even more quickly, giving an operating range of 75 miles, but on arduous cross-country going the figure could be as low as 60 miles (i.e. 0.5mpg!).

Given the many changes subsequently made to the design, including new guns and engines, it is worth mentioning the quality of its packaging. The armoured envelope was the thing that remained the same despite all those later modifications, and it was a huge improvement over previous British designs: the Centurion's front hull (or glacis) was sloped rather than flat; its belly was shaped like a boat's keel, deflecting mine blast upwards and outwards; and the turret's sides also sloped, notably between the thick mantlet mounting the gun and the roof.

This last feature, the turret front, was particularly important for tank fighting because, by giving more headroom for the gun's breech (the part that extended inside the vehicle) it allowed the weapon to be depressed further. In this way, the vehicle could take up a firing position on a ridgeline while showing very little of itself to the enemy. By way of comparison, the Centurion's 20-pounder could be depressed by 12 degrees, the guns of later Soviet tanks by just 2–5 degrees.

Emerging then from trials in the summer of 1945 was a vehicle that scored highly on protection and firepower. Its arrangement of power train and running gear also gave it a good cross-country speed, even if on-road it struggled to get past 25mph. The need for frequent refuelling was a problem but, as we will see, various workarounds would emerge for that.

As ministers pondered whether to greenlight production, there were some bigger questions. With the Pacific War now over, Britain and the USA had begun a major demobilization. The public yearned for normality and the rapid return of armaments factories to peaceful production.

Questions about the future need for the armed forces soon became focused on the Red Army. In the early months after the war, British ministers justified the continued existence of large armed forces by the need to contribute to the Allied occupation of Germany, Austria and Japan. Very quickly the mood of celebration evolved into one of suspicion and mistrust, with the Western Allies feeling that their continued armed presence was required to prevent the Soviet Union from taking more of Europe.

By early 1946, Winston Churchill would describe the new division of the Continent as an Iron Curtain. And a little over two years later, actual conflict loomed when the Soviets blockaded the Western Allies in Berlin, spurring the formation in 1949 of the North Atlantic Treaty Organization, Nato. These events marked the start of the Cold War, and in turn led to the definition of Western military needs in terms of the 'Soviet threat'.

Speaking to the House of Commons in 1950, for example, the Minister of Defence, Manny Shinwell, drew attention to the Kremlin's failure to demobilize, noting that they still had 2.8 million people under arms and spent the equivalent of 13 per cent of their economic output on the military. 'The existence of these vast forces', he went on, 'in the hands of a totalitarian State, where the pressure of public opinion does not operate and whose intentions are uncertain, represents a potential danger of which other nations must take full account.'

Just as the presence of these armies, at high readiness, would shape Western security in a broad sense for the next four decades, so the composition of those Soviet forces would drive the development of new weapons. Matthew Ridgway, the American general given supreme command of Allied forces in Western Europe, warned in 1952 of 'the mass of Soviet armor which will inevitably be launched against us at the very outset of hostilities'.

Did the fact that the Soviet Union kept large tank armies require members of Nato to do the same? It took time for a consensus to emerge. While Britain pressed on with the production of the Centurion, across the Atlantic they weren't quite so sure. For a period during 1946–47, tank production in America stopped. There was a feeling that there might be better solutions.

Late in the war the hand-held anti-tank weapon had evolved to the point that a new shell for the US bazooka could penetrate 280mm of armour. Again we see how the shaped charge, that detonation of explosive around a conical metal insert designed to produce a penetrating bolt of great power, posed existential questions for the tank. In the final battles for the Ruhr and Berlin, the Nazis had handed out tens of thousands of *Panzerfausts*, hand-held anti-tank rockets with similarly impressive performance. These weapons denied many a tank crew their dream of going home once the Reich had fallen.

Development of new lightweight artillery, dubbed 'recoilless weapons', meant powerful anti-armour weapons could be mounted on vehicles as small as jeeps. These guns were so-called because they vented almost the same blast out of their rear as came from the muzzle, effectively cancelling out the recoil. Add to that the rapid improvement of jet fighter aircraft, with associated weapons, able to threaten ground forces over wide areas. Anti-tank weapons had developed to the point where no tank, however heavily armoured, could withstand them, so why build more?

If the changed correlation of tactical forces, the battlefield factors, weren't bad enough for generals trained in the old ways, the advent of atomic weapons vaporized much of the old thinking about the higher levels of war also. How could forces – armoured or not – be massed to achieve a breakthrough, if doing this might create an irresistible target for a nuclear strike? And how could the national will to fight be maintained if cities could receive the Hiroshima treatment?

Speaking to the cadets of the US Military Academy at West Point in June 1950, the Army Secretary, Frank Pace, summed up the challenge: 'The principles of the recoilless weapon, the bazooka and the shape charge are being developed to a point where the mechanized-panzer blitzkrieg will play a much less decisive role than it did in the last war. Adding to those the more recent developments with regard to guided missiles and rockets, target-seeking equipment and the possibilities of tactical use of atomic weapons, it may well be that tank warfare as we have known it will soon be obsolete.'

Pace had read the writings of Professor Vannevar Bush, the

boffin who played a critical role in mobilizing America's scientific potential during the late conflict, and who in 1949 published a widely read book about the transformation of warfare in the nuclear age. The advent of so many low-cost anti-armour weapons meant, he believed, that 'a tank wandering through a country infested with such weapons would have a short life.' More broadly, he argued, 'there is a strong indication, therefore, that the defense may again be in the ascendant in land warfare [and] that the deadlock of the First World War might well reappear.'

Once the early US monopoly on nuclear weapons was lost, as happened later in 1949, Professor Bush predicted that warfare would still be possible, but would take a more dynamic and destructive form. In the United States these debates produced a belief early in the Cold War that strategic mobility was of the essence: marines and airborne forces would be vital, armoured formations less so.

Influenced by these ideas, Secretary Pace and the Pentagon hierarchy believed for a time that the answer to their tactical problem in Western Europe was not more tanks, but an armoured anti-tank vehicle called the Ontos. This curious-looking machine mounted six recoilless rifles, capable of firing a salvo of 106mm HEAT (High Explosive Anti-Tank or shaped charge) shells. It had a crew of just three, weighed less than 8 tonnes, and had broad tracks, all of which made it very mobile. The Pentagon at one point wanted to buy 10,000 of these new tank-killers. But in the end fewer than 300 were built.

What changed? To a large extent military conservatism exerted itself. The Korean War reminded generals that a tank had many uses besides killing other tanks. It brought scores of shells to fire in support of the infantry, whereas the Ontos carried just 18. That new vehicle was also too lightly armoured. The question of protecting a crew from the great majority of weapons that might be encountered on the battlefield reasserted itself: as the 8th Hussars on the Imjin had shown, heavy metal gave them the confidence to fight on.

Over time, as the USSR deployed more and more weapons capable of hitting America, ideas of 'massive retaliation' – that if Soviet tank divisions rolled west, Nato would quickly resort to large-scale nuclear warfare – became less attractive to the US. The Western alliance needed to be able to hold the line for longer without resorting to nuclear weapons and give itself more options, a doctrine that later became known as 'flexible response'.

Chiming in to these debates about future tactics, who should reappear but those 1930s British tank advocates Basil Liddell Hart and J. F. C. Fuller. Rehabilitated after his involvement with fascism, Boney Fuller wrote prolifically during the 1950s, for example telling readers of the *Royal Armoured Corps Journal* that atomic weapons strengthened the role of the tank. They would 'enhance the value of mobility, because rapid dispersions and concentrations, such as can be effected with cross-country vehicles, will become doubly necessary'.

Over time, the survivability of heavy armour emerged as a key advantage in this new age. In October 1953, as part of weapons effects trials, a Centurion was parked 320 metres away from Ground Zero during a British nuclear test in Australia. Its engine was left running and a crew of mannequins with radiation measuring devices left at the controls. After a blast estimated at 3 kilotons, scientists returned to the scene. The vehicle's engine was still running, and it was driven away. Analysis of the radiation badges showed that the crew would have died a few days later. However, the Centurion had shown it could operate on a nuclear battlefield, and Soviet and Western scientists went to work on 'radiation liners' fitted to the interior of their tanks, considerably improving the crew's protection. The tank used in the test was later shipped to Vietnam as part of the Australian military force there. Having survived an atomic bomb, in 1969 it remained in operation despite being hit by a rocket-propelled grenade.

Returning to the early 1950s, the nuclear test episode buttressed the Cent's reputation for toughness. Within a few years of the Second World War ending, the continued need for tanks had been accepted by many armies. Britain's decision to press ahead with the Centurion had, accidentally or not, conferred a kind of leadership not seen since 1916. Countries around the world looking to replace worn-out Shermans, T-34s or indeed Panzer IVs had very few options. Only the USSR, UK and US were making something more modern.

Perhaps it is a typically British parable that the birth of this tank was surrounded by so much argument and confusion. Researching the minutes of the Tank Board, Ministry of Supply or General Staff, you read a bewildering variety of ideas about what they were trying to achieve: 'universal', 'capital' or 'victory' tank; 'heavy cruiser'; 'British Panther'; 'British Tiger'; and 'heavy Cromwell' are all terms that crop up.

Ultimately, the best expression of their ambition for the vehicle comes in a minute from a 1947 meeting: the Centurion was intended to be 'a match for any tank in any army'. And undoubtedly it succeeded by that measure.

The new machine became an enormous export success. During the 15 years following the war, 15 countries bought the Centurion, contributing £130 million to Britain's public finances. Customers ranged from Nato nations like Canada and Denmark to the non-aligned Sweden and Switzerland, India too, and countries across the Middle East. It was Israel, which put its first Cents into service in 1960, that would be the tank's largest user in battle.

It wasn't just that some of these armies lacked alternative suppliers. The Centurion's export success derived from the quality of its design, its clear technological superiority (for example, of the Mark III version used in Korea over the Sherman), and its adaptability. With its long development and rising price it became an archetype of the Cold War arms race.

While the Centurion had been developed initially at a wartime pace, with prototypes arriving in barely 18 months, the end of the war slowed everything down. There were upsides to that. Information from trials in Germany was combined with the recent battle experience of so many of those connected with the project to ensure that the vehicle embodied many innovations.

The Centurion was the first British tank to be fitted with a full gun stabilization system. As the tank moved, spinning gyros measured its pitching and rolling, with the control system then keeping the gun level by means of electric motors. That meant that all but the most extreme jolts on the vehicle could be ironed out, and the armament kept locked on its target while motoring along. Although the Sherman and late-model T-34s had rudimentary systems to do this in one axis (elevation, i.e. moving the gun up and down), the new British machine had it in traverse too. Firing on the move suddenly became far more accurate.

In order to power this new gun equipment, as well as radios and other systems, a small auxiliary engine was fitted. So crews sitting for long hours in one position could run this Morris motor to keep the vehicle powered up, while switching off the very thirsty Meteor main engine.

The Centurion also contained many 'soldier-friendly' innovations such as the boiling vessel (effectively a militarized kettle) that meant they need never run short of hot tea. It had a phone on the back that allowed infantry to communicate with the crew, something that had been tried as a modification to previous tanks but here was built in from the start. Its fire suppression system was also a step ahead of wartime extinguishers, and there was an escape hatch in the back of the turret. These reflections of recent battle experience made it more popular still with crews.

Of course, all of this came at a price, as did the relatively slow production rates. Estimates of a weapon's cost are notoriously difficult, not least because of the different definitions of what was included in contract prices. Pre-production estimates of £21,000 per tank can be found in army papers. However, the figures given to Parliament showed a climb in cost from £35,000 early on in its production to £50,000 in 1949.

By way of comparison, the Matilda II infantry tank had come in at £18,000 in 1940. Where the Detroit system of mass production had been used, costs were considerably lower: Britain's Shermans were priced at $35,000–40,000, equating to £8,750–10,000 per tank at wartime exchange rates. The Centurion was therefore four or five times more expensive than vehicles bought just five years earlier.

In one sense this was just the price of technological advancement. The gun stabilization system alone cost £1,600 per vehicle, for example. A Centurion had around 30,000 component parts compared to the M4A4 Sherman's 4,537. As for the speed of production, a skilled labour force required four months to build a Cent, and the army's initial hopes of making 40 a month took years to materialize. This was just a fraction of wartime rates, inevitably adding costs.

In all, 4,442 Centurions would be made between 1946 and 1962. Early production was focused at the Royal Ordnance Factory Barnbow in Leeds, with another plant later coming on-line at Leyland in Lancashire. Production peaked at 573 tanks during the 1952–53 financial year. These might be small numbers compared to wartime, but they were very substantial compared to more recent times.

That the vehicle's production and service continued for as long as they did owes much to its adaptability and the nature of Cold War military competition.

The tumultuous events of October 1956 in Hungary were to spur a fresh round of developments in Western armoured forces. A decision by the Soviet leadership to crush a patriotic revolt brought street battles to Budapest and other places, pitting Hungarian patriots against Soviet tanks. It was a Cold War watershed moment in many ways.

The Kremlin's willingness to use force on a large scale against those seeking greater freedom was emphatically demonstrated. Images of armour rumbling through the streets reinforced ugly associations of these machines as behemoths used to crush or overawe those with the most human of demands. The Hungarian revolt also allowed Western countries to gain a greater understanding of how the Soviet army had evolved in the 11 years since it took Berlin.

For Lieutenant Colonel James Cowley, the British Military Attaché in Budapest, this crisis presented opportunities as well as risks. When, on the 25th of October, his driver, László Regéczy-Nagy, spotted two abandoned Soviet tanks outside a barracks where a rebel Hungarian colonel had repelled the Russians, he could see his chance. 'I at once stepped on the accelerator to report the finding to the colonel,' Regéczy-Nagy later recounted. These two machines featured in some photos of the day's street battle. In one, triumphant Hungarians hold their flag aloft over the tank, while people mill about in the street. Returning with his driver, Lieutenant Colonel Cowley instantly understood that these trophies had an altogether greater significance even than the jubilant crowds realized. Cowley, who had commanded a squadron of Shermans on D-Day, realized that the rebels had captured the Soviet Union's most advanced tank, the T-54.

This machine was a product of the Kharkov design bureau run by Alexander Morozov, the same team that had produced the T-34. In many ways it was an evolutionary development of that wartime tank, having a very similar hull design, mated with a well-rounded turret (occasionally compared to an inverted saucepan) and a powerful 100mm gun. The T-54 and its younger brother, T-55, would be produced in huge numbers, becoming ubiquitous Cold War fighting machines. At this moment in 1956, though, they were brand-new, and only just entering service.

Cowley realized that it was a moment full of risk. For one thing, Russian soldiers were trying to regain the initiative, the streets around echoing with gunfire as they approached. He could easily get shot. Scurrying from doorway to alley, the British officer was able to escape. He hatched a plan with his driver who, as luck would have it, had served as a tank driver in the Hungarian army.

That night Regéczy-Nagy returned to one of the T-54s, jumped in the driver's hatch and started it up. Rumour had it that Cowley had done a deal with the rebels, providing them with information in return for the tank. It was driven into the British embassy compound, where for two days Cowley and his staff made a detailed inspection, took photographs, and generally satisfied their curiosity. It was then returned to the rebels, who fought on into November but whose cause was ultimately doomed. The Hungarian revolt was crushed with the loss of thousands of lives. Those on the left in Hungary who despaired at this labelled former comrades who sought to justify it as 'tankies'.

Going back to Lieutenant Colonel Cowley, his windfall had proved one of the highlights in the decades-long East–West intelligence battle. So often the analysis of a new Soviet weapon depended on educated guesswork, literally people poring over photos making estimates, following its appearance in parades or on manoeuvres. The British attaché had seized the chance to deliver a much fuller appraisal.

At Chertsey and elsewhere, Western boffins digested unsettling news with implications for the Centurion's gun and armour. While the 20-pounder had very good performances against flat plates (being able to penetrate 290mm of hardened steel even at 2,000 yards (1,800m), it slipped markedly against sloped surfaces. Indeed, it was reckoned that against a plate raked at a 60-degree angle, even at half that distance it could only penetrate 87mm. They now knew in the West that the T-54's armour was 100mm thick.

So the Royal Ordnance Factory was tasked with fielding a gun quickly that could defeat it. Their answer, and it was already under development before the Hungarian rising, was essentially their 20-pounder (or 84mm) gun design rebored to 105mm. Called the L7, it was an even bigger success than the Centurion itself, becoming the standard armament of most Nato tanks. The new 105mm gun, firing Armour Piercing Discarding Sabot shot (the type where a casing around the penetrator flies off shortly after firing), could penetrate armour around 50 per cent thicker when sloped at 60 degrees. It was comfortably enough to defeat the T-54, with subsequent ammunition developments keeping it effective for many years.

British scientists had their worries about the Soviet heavy tanks even before the events in Hungary, and so had developed a British equivalent, the Conqueror, armed with a mighty 120mm gun. The Conqueror served in small numbers for just 11 years (1955–66), and its only significance in this story is that its existence represented a last gasp of Second World War heavy tank concepts, and denied the Centurion the distinction of being the world's first 'universal' or 'capital' tank.

Meanwhile, improvements were incorporated into the Cent. Its frontal armour was upped to 126mm on the hull, built into the Mark VIII. By 1959 new Centurions were incorporating both the thicker armour and the 105mm gun, with a programme to upgrade older marks too.

If this saga serves as a parable of the Cold War arms race, the unending jockeying for advantage between two hostile blocs, we should also remember that it took place on many different fronts. Each side had its proxies, notably in the Middle East, where the balance of nuclear terror did not prevail, allowing for warfare to continue.

Just after the Hungarian uprising, British troops invaded the Suez Canal zone in Egypt. Centurions of the 6th Royal Tank Regiment took part, and their short engagement against a troop of Egyptian army SU-100 tank destroyers marked the only occasion apart from that single time in Korea when British army Centurion crews engaged enemy armour.

The Middle East, however, was to prove the arena where this tank really proved its worth. In an echo of the freewheeling attitude to arms sales of an earlier age, Britain sold Centurions to Egypt, Kuwait, Iraq, Jordan and Israel.

It was the Israelis who used them to greatest effect, augmenting their fleet with dozens of captured Egyptian, Jordanian and Iraqi Cents. And it was Israeli commanders who, during the 1967 Six Day War, reminded the world of the unique suitability of the desert for armoured warfare, emphatically reasserting the continued value of the tank. Indeed, these events were to mark a swinging back of the pendulum towards the tank and away from all the many defensive weapons designed to destroy it.

Integral to that story were Centurions and a former British army sergeant who knew best how to use them.

In the long-running struggle between Israel and its neighbours, the affair of the 3rd of November 1964 was a minor skirmish. In the story of the Centurion tank, however, it was altogether more significant. It took place on the border with Syria, at a critical point just south of the Golan Heights where the River Jordan snakes south.

At the time, Israel had recently opened a national water carrier, diverting much of the river's flow westwards, towards agricultural projects and towns in its interior. Syria decided to launch its own project upstream of that canal, so that it might bleed off the stream to irrigate Arab lands instead.

In the Nukheila area, where the clash took place, the Syrians regularly shelled *kibbutzim* (collective farms) near the frontier, taking advantage of the observation posts it had on the Golan massif. For their part, the Israelis used these clashes to take potshots at the bulldozers and other equipment working on the Syrian side. On the 3rd of November, the Israeli Defence Forces (IDF) had a platoon of four Centurion tanks there. The Syrians had a similar number of Panzer IVs, which used revetments or prepared firing positions to screen themselves.

Shortly after noon, a Syrian machine gunner opened fire on Israeli troops patrolling near the border. It was time for Israeli Centurions to make their combat debut. Moving forward from their hide positions they began engaging Syrian armour, anti-tank guns and infantry positions. The Syrians stepped it up too, bringing accurate mortar fire onto the IDF tanks, making the crews close down.

After 90 minutes of tit-for-tat, United Nations observers

managed to halt the firing. Hours later, Major General Israel Tal, appointed two days earlier as the commander of the Armoured Corps, travelled up to debrief the crews. Freshly promoted, his short stature led to the nickname 'Talik' or 'Little Tal', but he was to have an outsize role both in the Centurion story and that of the IDF's tank arm.

Reaching the scene of the recent engagement, he was in turn bewildered then angry. The four-tank Centurion platoon had fired 89 shells from their main guns but not managed to hit either of the Panzer IVs that engaged them. Dust from their own firing, Syrian suppressive fire and poor crew procedures sighting the guns had created a debacle. 'The first results with the newly acquired British tanks had been disappointing,' wrote Lieutenant Colonel David Eshel, an IDF armour veteran and later historian, arguing that the maintenance regime was 'much too complex for the inexperienced tank crewmen'.

In desert exercises the Cents often broke down because of sand ingestion. The Israelis regarded the Sherman as much more reliable. And when it came to their guns, bore sighting the British gun – that is, making daily checks to ensure that the barrel remained aligned with the sights – seemed to be beyond the ken of many Israeli crew. Tal was furious. He regarded the crews' failures as symptoms of deeper discipline and professionalism problems which he had to address. In the IDF's jumble of nationalities and military backgrounds there were tank soldiers with US, French and Soviet training. So he was determined to prove the Centurion's value.

Tal, who had been born not far from the scene of the recent skirmish 40 years earlier, was an Anglophile. During the Second World War, he had served as a sergeant in the British army's Palestine-recruited Jewish Brigade. Unusually, then, he was both a Jewish intellectual and someone whose ideas on discipline followed British army lines. He was small in stature but large in intellect, and he certainly didn't suffer fools.

After ordering an inquiry into the events of the 3rd of November, Tal summoned all armoured unit commanders for a dressing down and a bracing explanation of the professional standards he expected them to achieve. It ran from everyone wearing the same coloured socks to enforcing rigorous daily vehicle checks by crews.

Just ten days after the first Nukheila incident, trouble flared again on the Syrian border. This time, when the foot patrol came under fire, there was an armoured halftrack ready to pick them up. A second platoon of tanks, trusty Shermans, had also been assigned to support them, and was soon ordered up. One of the drivers made a mistake, though, and his vehicle careered off the embanked road it had been using. Quickly, a pair of Centurions moved up, engaging the Syrian tanks. This time, as you'd expect in such an unequal contest, they quickly knocked out a pair of Panzer IVs. Things escalated from there, the Syrians increasing shelling of the nearby kibbutz, and the Israelis ending the affair with an airstrike on the artillery.

Flare-ups continued along the border during the following months. IDF Centurion crews knocked out Panzer IVs, T-34/85s and SU-100s, many of them at ranges of 2,500–3,000 metres, well beyond those the Syrian vehicles could manage. The SU-100, a Soviet-made tank destroyer with steeply sloped armour and the same gun as the T-54, had been encountered by the British during the 1956 Suez operation. To meet these emerging threats, the Israelis followed the British lead, converting their Centurions to the 105mm gun. They also started welding an external fuel tank to the hull rear to boost its range.

On the 12th of August 1965, nine months after the initial Nukheila incident, there was another tank fight around 40 miles to the north. It also involved Syrian attempts to divert a watercourse, this time a good deal further away from the border. Like the earlier skirmishes it began with small arms fire, with the Centurions motoring from hides to their firing positions.

During a running battle of three hours, the IDF armour, fitted with the new 105mm guns, knocked out four tanks, and some tractors nearly 7 miles away. One of the Cents suffered a direct hit on its roof from a Syrian artillery or tank shell with no ill effect to the crew – further enhancing its reputation. Emerging from one of the tanks at the end of it was none other than Talik, boss of the Armoured Corps. Dismissing objections, he had insisted on taking the gunner's seat, personally knocking out a Syrian tank. Lieutenant Colonel Eshel wrote: 'The Israeli tank units systematically destroyed every piece of Syrian equipment . . . The Centurion was vindicated.'

As for Tal, he pressed on with his plans to raise standards, while introducing more Centurions across the force. Many in the Royal Armoured Corps and British arms industry relished close cooperation with Israel, since generous Soviet backing for Egypt and Syria made it a proxy forum for Cold War armoured developments. The Foreign Office, on the other hand, worried about its relations with the Arab world. Hence Centurion deliveries to Israel went slowly and were balanced with those to Western-friendly Arab neighbours.

By the summer of 1967, though, Israel had built up a force of more than 250 Centurions, making it the backbone of Tal's armoured corps. Furthermore, they were offering to make a big investment in the British army's Centurion replacement, the Chieftain. These plans were disrupted by what happened next.

When tensions between Israel and its Arab neighbours finally produced a regional war, Tal, the ex-British army sergeant, found himself in command of the 84th Armoured Division, striking against the Egyptian army in Sinai. He would therefore join Guderian and Patton in that select group who had both developed armoured forces and commanded them in battle.

Having detected a big build-up of Egyptian forces, Israel unleashed a devastating pre-emptive strike. Its air force destroyed scores of enemy planes on the ground, then three armoured divisions were unleashed into the desert.

In a speech to his commanders on the eve of battle, Tal echoed some of Patton's language from the Normandy breakout, telling them not to worry about their flanks, to press on relentlessly. His address was widely quoted in the IDF later. There was 'no alternative', he insisted, and success was vital: 'If we fail in the initial clash, the nation will be overrun. The fate of the nation rests with what we in our division do tomorrow. The survival of our country depends on us . . . Each man will assault to the end, taking no account of casualties.'

On they rolled. Tal's 84th was responsible for taking the Gaza Strip, before moving towards the Suez Canal. Khan Yunis and Rafah Junction, places so familiar from recent headlines, were captured in two days before he continued his advance. The IDF defeated an Egyptian force of 100,000 troops with 950 tanks and 1,000 artillery pieces, advancing well over 100 miles in only a few days.

Among the engagements they fought – keenly studied by Nato armies in the months that followed – were tank duels against the Soviet-made Stalin heavies, as well as brand-new T-54s and T-55s recently delivered to Egypt. In many of these, Tal's techniques of long-range gunnery, exploiting the capabilities of the L7 105mm gun, and of the Centurion more generally, won the day.

It did not prove an entirely positive experience from the Israeli tankers' point of view. On the third day of the war, as it moved to take the Mitla Pass through the highlands at the centre of Sinai, most of one Centurion battalion ground to a halt, having run out of fuel.

Soon after moving against Egypt, the IDF opened new fronts. It attacked the Jordanians in the West Bank and, further north, took the Golan Heights. The positions used for years to shell the Israeli kibbutzim in the Galilee had been wrested from Syrian hands.

The Six Day War was a victory on an epic scale; indeed the British military thinker Basil Liddell Hart described it as 'a perfect blitzkrieg'. The idea of great conquests by mechanized armies was reborn.

Of course, this success had other meanings too. It was the start of an occupation (particularly of the West Bank and Gaza) that would embroil Israel in perpetual controversy, denying it the security its operations were intended to achieve. The legal and moral consequences of ruling millions of Palestinians are beyond the scope of this book, but the political implications would prove highly significant too, because the victory would affect Israel's access to foreign military technology, including tanks.

Until the Six Day War, Israel had bought its armour from the French, British and Americans. Calls for a withdrawal from the recently occupied territories, coupled with commercial reprisals against certain Western countries, changed this. France was quick to freeze arms supplies; Britain took a more complex line.

Major General Tal and the IDF's other leaders, delighted with the Centurion's performance, placed an order for hundreds more, and in November 1968, looking to the future, asked to buy 250 Chieftains. British Cabinet ministers wrestled with this before blocking the sale one year later.

The Foreign Office view was that such a sale would ignite fury across the Middle East. Denis Healey, the Defence Secretary, felt it would 'put in jeopardy our future arms sales to the Arab countries'. Roy Mason, president of the Board of Trade, saw it even more starkly: 'the effect on the Arab world of the sale of Chieftains to Israel would be so great on our economy that it would knock us back for two years.'

So, two Chieftains that had been sent for trials in Israel were returned. Britain, however, continued to sell them Centurions, covertly delivering ten British army surplus vehicles each month. By the time of the 1973 Yom Kippur War, the Israeli fleet had expanded to 700.

As the question of future deliveries played out, the IDF drew its own conclusions from the Six Day War. Its tank repair workshops at Tel Hashomer, not far from Tel Aviv's international airport, were tasked with a major programme of upgrades to the Centurion. It was growing into a tank factory. Vehicles not yet equipped with the 105mm gun had it installed, and the decision was made to ditch the Meteor engine. Rebuilding the rear of the vehicle, the Israelis installed a US-made Continental diesel.

There were many advantages to this: it was air- rather than water-cooled; did not run on petrol; produced significantly more power than the British plant (730hp compared to 650hp); and had the logistical benefits of being used on the IDF's American-supplied M48 Patton tanks too. Critically also, once the new engine and additional fuel storage had been fitted, it allowed a near doubling of the Centurion's operational range.

The Six Day War marked a turning point in so many ways, but seen from the military perspective it saw the US stepping in as a major arms supplier, with the eclipse of the UK and France. In lieu of the Chieftain, America promised the IDF its new frontline machine, the M60.

For General Tal, smarting after being denied the next-generation British tank, this outcome was unacceptable. The Chieftain represented a significant improvement in armoured protection and firepower – it was markedly better than the M60 in these respects. Tal resolved that if the IDF couldn't have it, they would have to come up with their own answer.

T-64

Entered service: 1966
Number produced: 13,000
Weight: 38 tonnes
Crew: 3
Main armament: 125mm gun
Cost: R 190,000 (in 1969, equivalent to £105,000, or £1.48m in 2024 prices, but like the T-34 there is a sackful of caveats relating to the fact that the Ruble didn't trade freely and tank inflation is not the same as the consumer kind)

T-64

February 2015 in the Donbas, eastern Ukraine. As international moves to gain a ceasefire picked up, the protagonists struggled to gain a late advantage before the battle-lines were frozen. The Russians, using a pincer movement, strove to trap Ukrainian troops in the town of Debaltseve, while they tried to prevent that happening.

Meeting in snow-covered fields near a town called Logvinovo, on the last main road that the defenders were able to use, were two forces of tanks. The numbers, especially in terms of what would happen here seven years later, were trivial, but the symbolism deep.

Ukrainian Lieutenant Vasil Bozhok, as the fighting got underway early on the 12th of February, reckoning it would be a difficult day, took the gunner's seat in his T-64. Speed of engagement would be vital, and he wasn't quite sure that Vitaly, a conscript who usually sat there, would be up to it.

It was Bozhok's second battle. A couple of weeks earlier he'd been fighting nearby, having the satisfaction of knocking out a T-72. He'd also experienced being hit and damaged, having to leave the battle with a jammed autoloader and smoke filling the turret.

In this second engagement Bozhok's replacement was one of eight Ukrainian tanks facing the threat to the east, as a dozen Russian armoured vehicles, churning through the snow, approached. It was a competently conducted move, so much so that the Ukrainians reckoned they were up against Wagner Group mercenaries, though there is also evidence that it was a regular Russian army unit sent into the Donbas covertly, because at that stage the Kremlin was claiming its separatist proxies there were doing all the fighting.

Either way, the battle came on quickly. The Ukrainians had the advantage of being static and using the ground to cover themselves. Bozhok squeezed the trigger and let fly a 125mm shell: 'we immediately opened fire. The first shot was successful.' A Russian T-72 suffered an internal explosion, blowing off its turret. Bozhok's tank pulled back into cover as the automatic loader whirred and clanked, placing another shell in the breech. 'We also managed to knock out the second tank.' After firing two rounds, 'it stopped and never moved again.' Knowing that shells were coming the other way too, he used a couple of knocked-out vehicles as cover while he reloaded again.

On each side there was smoke rising from knocked-out vehicles. It was a symbolic contest because the T-64s had been made in Kharkiv, a Ukrainian city (called Kharkov in Russian), at the factory where the T-34 was born, and the T-72 built in Russia. Each army's propagandists extolled the superiority of their own machine.

At a range below 500 metres, Bozhok drew a bead on a third T-72, letting fly as it crossed a bridge just to their front. Whether through the force of the hit or a panicked driver, it tipped off the bridge into an icy stream. Bozhok had killed three T-72s in less than 20 minutes.

Ukrainian paratroopers, withdrawing from Debaltseve, started moving past the tanks. 'The situation was difficult,' Bozhok explained. 'They were running out of ammunition. We gave them a chance to regroup and get away safely.'

When he was down to his last two main armament rounds (he kept them, 'just in case'), the lieutenant quit the battle-line and headed north to join up with the other Ukrainian units. Captain Oleksandr Netrebko, commanding one of the units that was falling back, attested: 'we were saved by one tank crew that went into action and destroyed three tanks and a BMP [infantry fighting vehicle].'

The Ukrainians would claim eight T-72s knocked out in all, compared to four T-64s lost on their own side. For his part in this action, Bozhok was awarded his country's highest decoration, the Hero of Ukraine medal. Overall, the outcome at Debaltseve was a significant setback for his side, so his story may fall into that category of an army finding heroes to put a positive gloss on a defeat.

However, in the context of tank history, the action emphasizes what a remarkable tank the T-64 is. It had entered service nearly 50 years before Bozhok's engagement and yet remained capable of beating a more modern machine.

At the time of its birth, half a century earlier, the T-64 was a bold leap forward. Unsurprisingly, perhaps, this was a story with plenty of setbacks and its significance was poorly understood at the time in Western countries. But of all the vehicles in this book, the T-64 was the one where the broadest range of new technologies was introduced simultaneously, a truly revolutionary development at a time of intense Cold War competition.

Back in 1942, the Kharkov Locomotive Factory, scene of the T-34's development and initial production, had been occupied by the Nazis and its design teams displaced to the Urals, but by June 1945 it was back in production. In the 1950s it would be named after Stalin's troubleshooter Malyshev, the wartime armaments boss, who had said: 'we cannot crush them with numbers . . . quality is the path to supremacy.' And post-war the design team there certainly embraced that philosophy.

Alexander Morozov, who had continued development of wartime vehicles after the death of Mikhail Koshkin, was firmly in charge. Sharing the laurels for the creation of the T-34, he had been garlanded with almost every honour the Soviet state could heap on him, from the Hero of Socialist Labour medal to state prizes and a major general's rank. The Morozov design team did not rest, producing the T-44 in an attempt to keep up with late-war German tanks, and then, soon after the war ended, the T-54.

In common with Western designs, these machines continued the arms race, mounting heavier guns as well as having thicker armour than those they were meant to replace. Soviet designers, though, did not allow weight and size to creep up too high, not as far as their medium tanks were concerned, in any case.

In the immediate post-war years, the task of developing advanced armaments became a good deal harder. Greater sophistication led to longer development times and rising costs. These dynamics played out a little differently on the two sides of the Iron Curtain.

In the Soviet Union, the crazy energy of the war years, trying to outproduce and outsmart the invader, was gone. Stalin's terror had given way to a period of 'collective leadership'; purges were out of vogue. And in the myriad research institutes, engineering shops and design bureaus, a great many sought to justify their existence and the continuation of a privileged lifestyle, leading to foot-dragging or overt hostility between them. These squabbles and the competition for resources were catalogued in a diary kept by Morozov that was published after his death. It includes rough minutes of the high-level Communist Party meetings where officials tried to reconcile competition and favour those most likely to deliver the best weaponry.

Evidently Morozov felt the Cold War rivalry keenly. In January 1952, for example, having received a briefing from military intelligence on the new US M47 tank, he wrote: 'the Americans do not stand still and follow our path. We need a breakthrough and urgently.' Buoyed by his reputation as a creator of the T-34, Morozov was convinced that a similar bold advance was needed. He sought simultaneously to combine the qualities of a medium tank with the power of a bigger beast.

The early post-war Soviet army did continue with the development of heavy tanks, the T-10 succeeding the Stalin or JS-III. The T-10 packed a formidable punch, in the shape of a 122mm gun, and was well armoured. It was only by giving the crew minimal space inside that they kept the weight of the T-10 to 49 tonnes, and even this was too heavy in the view of many Russian generals.

It was open to question whether the investment was worth it. If a hand-held anti-tank weapon could pierce any tank available in the 1950s with its shaped charge warhead, what was the point in piling on more and more armour, with all of the mobility challenges that would bring? For British designers, the attempt to protect the crew from as many threats as possible, particularly tank guns, meant accepting that the 50-ton Centurion would be replaced by the 57-ton Chieftain.

But for the Soviet state committee directing the tank industry, this was unacceptable. In 1961, they issued a specification for a new medium tank of ambitious performance and firepower that should weigh just 34 tonnes. A gauntlet had been thrown down: how would the design bureaus meet it?

At Nizhny Tagil in the Ural mountains, the office originally set up by the Kharkov evacuees had become an independent entity when they returned home at the end of the war. Once endowed with its own design bureau, Nizhny Tagil's formidable production line would compete against its wartime bosses for orders. Plant directors in the USSR may not have had the same rewards that capitalist ones did, but they did stand to gain personally (for example, with state prizes or sinecures), spawning an intense rivalry. The Urals-based bureau had won orders for thousands of T-62s, very similar in appearance and weight to its predecessor, but mounting a bigger gun (115mm).

Morozov rejected that approach, instead re-examining every aspect of vehicle design. It could not be done by further evolution. As Morozov wrote to a senior officer: 'the solution to such a problem, you will not deny, is only possible if, to some extent, it is reasonable to break away from the generally accepted "canons" of tank building and choose new ways that would open up new opportunities.' Everything would have to be rethought. Ideas developed at the Kharkov design bureau from the mid-1950s onwards were embodied in a prototype, *Objekt* 432, which in turn became the T-64.

If a vehicle's weight, for any given amount of protection, was dictated largely by its volume, how could that be reduced? Installing an automatic loader, thus reducing the crew from four to three, was one answer. In the T-55, for example, the human loader had required 1.36 cubic metres of space. How else to reduce volume? Morozov had previously realized that mounting an engine transversely (i.e. side-on) could do that, and now he looked for an even smaller engine, with a simpler cooling system stacked on top of it, as a further saving. Continuing in the tradition of the T-34, you could also limit the fuel carried under armour by mounting extra tanks on the vehicle's exterior.

As for protection, by making the volume smaller, eliminating a loader who needed to work standing up, you could reduce the silhouette of the vehicle so it was a lesser target. The T-64 would stand 2.17m high rather than 2.4m for the T-54.

Research in secret Soviet scientific institutes had by the early 1960s produced other answers to the protection conundrum. They experimented with composite or laminar armour. Simply put, they wanted to know whether the bolt of metal produced by a shaped charge (or High Explosive Anti-Tank shell) could be countered more effectively by layers of different substances that might spread or reduce its energy rather than by a single, homogeneous plate of hardened steel.

T-64s were therefore built with layered armour on the hull front or glacis, as well as the forward parts of the turret. This sandwich of materials included hardened steel outer and backing plates with a type of fibreglass in between. As with the Kevlar fibres used in bulletproof vests, this type of structure helped to dissipate energy. Turrets were built with dozens of ceramic spheres filling the space between steel layers. These 70mm-diameter balls, made from a porcelain-type substance called *ultrafarfor* in Russian, were also found to be effective and lighter than steel. Among Western scientists working on similar technology, the key concept was termed 'mass efficiency'. It was about improving on the protection provided by rolled steel armour while doing so at a lighter weight.

Soviet tests in 1967 showed T-64 turrets could withstand direct hits from 100mm and 115mm HEAT and armour-piercing rounds. Soviet scientists were confident they had come up with armour that was impenetrable to the L7 105mm gun used by most Nato tanks.

These composites – whether fibreglass, porcelain, aluminium or other types of steel – could not negate the fact that defeating the latest Western weapons required very thick armour. The T-64's glacis was more than 200mm and its turret front or 'cheeks' exceeded 400mm. But it could incorporate these at a significantly lighter weight than earlier vehicles. To understand the achievements of Morozov's design, consider that the Tiger, with its 100mm hull front and 120mm armour around the gun, weighed in at 57 tonnes. The T-64, with two to four times the depth of frontal armour, was around 20 tonnes lighter. This was primarily achieved by reducing its internal volume, and secondarily by using composites.

Protection against weapons of mass destruction had to be built in as well. That meant lining the vehicle's interior with radiation-absorbent material. It also required fitting an air conditioning system powerful enough to create an overpressure inside the crew compartment, something that stopped chemical agents from dripping in through the gunsights or other small openings in the armour. Similar equipment was fitted to Western tanks – just one more aspect in which Cold War vehicles embraced greater and greater complexity.

In order to crush down further the vehicle's size, the Kharkov designers decided to use a different engine type. It took up markedly less space than the V12s in the T-34 or T-54 because it had just five cylinders, inside which ten horizontally opposed pistons operated. It was a design cribbed from a German aero engine developed in the 1930s. Here Morozov and his comrades overreached themselves, because this motor would give them no end of difficulties.

When it came to mobility more generally, they missed no chance to innovate. The tracks used a dual pin design with the links themselves being a 'snow shoe' shape in order to keep the ground pressure as low as possible on the spring and autumn mud so well known in that part of the world. The T-64's road wheels did not use the rubber tyres favoured on many other tanks – because of concern they would melt or split under certain circumstances, including nuclear war! Instead, vibration was ironed out by building shock absorption into the inside of each wheel.

As for firepower, the tank, having initially been armed with a 115mm cannon, was upgraded to a 125mm one – a calibre that remains the largest in service to this day. This uprated model, termed T-64A, packed enough punch to drill through the Chieftain's armour.

Increasingly by the 1960s, designers had realized that by lengthening the solid shot at the core of an anti-tank round, its armour-piercing performance could be boosted. It relied on the stiletto heel effect, or the way the tip on a knight's poleaxe pierced the breastplate of his opponent. This meant focusing to a point the energy of a very dense penetrator, fired at something like 1,800 metres per second muzzle velocity. That was logical but came with complications. Once the length of the penetrator exceeded its diameter by a ratio of more than four or five to one, the round became unstable, wobbling in flight. But by turning it into a dart, with fins on the tail, it could easily be doubled in length or made even longer. Like so many of the engineering solutions, addressing one issue created another.

It was difficult (though not impossible) to launch a fin-stabilized round from a gun barrel that was rifled or grooved. Rifling of everything from bullets to tank rounds is done to improve accuracy by imparting spin. But fins got in the way of that. The Soviets therefore continued the practice begun with the T-62, giving the T-64 a smoothbore (i.e. unrifled) barrel (initially the same 115mm weapon as the T-62, but switching quickly to the T-64A with a 125mm gun with an even heftier punch).

As so often in this story, the adoption of one solution had effects in other areas. Western rifled guns like the British 105mm and 120mm were better for accurate long-range shooting. But at an early stage in developing his new tank, Morozov – who knew that most tank battles in the late war had taken place at hundreds rather than thousands of metres' distance – argued that this was less important.

In his diary he wrote early in 1961: 'we must abandon tank duels at long distances. These are the tasks of ground-based systems and aviation. A tank is a melee weapon.' This decision had big implications. As we've seen, Israeli crews trained to engage at 2,000–3,000 metres, as did many Nato crews. By designing their tanks to focus on closer ranges, the Soviets ensured they would often be outranged. All Soviet tanks of the later Cold War period adopted the 125mm smoothbore gun, which performed best below 2,000m. The consequences would be felt from the deserts of the Middle East to the plains of the Donbas.

Of all the breaks with convention embraced by the Kharkov design bureau, arguably the fitting of an autoloader was the greatest. Twenty-eight projectiles, with their separate propellant charges, were stored in a carousel on the turret floor (with eight more elsewhere). After each round fired, it revolved clockwise, allowing a mechanical arm to move a fresh shell or shot up into the gun breech, followed by its charge.

With different ammunition types, the next round in the carousel was not always the kind selected by the gunner, in which case it would continue to turn until the right one was aligned. This meant the reload speed could vary from six to 20 seconds, depending on where the next round was. While this was fine at the lower limit, a trained human loader could get several rounds off in a minute.

By the mid-1960s the systems used for aiming guns were evolving too. Like the Centurion, the T-64's gun was fully stabilized. As the T-64 evolved it was later fitted with a fire control system – a computer that allowed the gunner to 'pickle' a target by placing an aiming point in the sight on it. The processor would then make all the calculations necessary to hit it, factoring in everything from the tank's movement to the wind speed.

Summing up all of these innovations, Morozov evidently was determined to buck the trends that had overtaken tank builders during the preceding decades, particularly in the West, of an inexorable growth in size and weight. In doing this he achieved that elusive goal, cause of so much debate in Britain's Tank Board, of creating a 'universal' tank.

It is unsurprising that, having embraced so many different new ideas, the Kharkov team should have encountered myriad challenges turning the Objekt 432 concept into a workable vehicle. For three years in the early 1960s they tried to overcome them. Early in 1964 matters came to a head. Morozov and other key defence industry officials were summoned to Communist Party headquarters in Moscow for a two-day conference. Such encounters, where portly, grey-suited bosses, often smoking away, berated underlings who had come up from the provinces, were well-established rituals. Inevitably, they would try to determine why things were going so slowly before castigating the slackers.

The bosses eventually agreed to sign off on Morozov's continued work, but only after he had to listen to a litany of complaints about it. Among those who vented their frustration was Pavel Rotmistrov, the commander who had sent his 5th Guards Tank Army to its destruction at Prokhorovka 21 years before. Having escaped being purged by Stalin, Rotmistrov had been promoted to chief marshal of tank troops, attending the 6th of February meeting as the main representative of the 'customer', if such a concept is not too far removed from the Soviet way of doing things.

Rotmistrov told the assembled apparatchiks that he accepted the basic concept of the Kharkov design, noting: 'we need a small-sized tank – that's why we are attracted to the [Objekt] 432 in terms of its shape, but we need to achieve the best.' He then cast doubt about whether the new compact engine could come up to scratch, remarking rather presciently that when it came to powering increasingly heavy armour, a gas turbine was 'the engine of the future'.

This idea would cause Morozov considerable grief, because it undermined confidence in his compact diesel and led eventually to the development (in the rival Leningrad plant) of a gas turbine-powered version of his vision, later dubbed the T-80. Work continued through 1964, as engineers battled to solve myriad problems with the Kharkov design, including those with its powerpack.

When the party bigwigs met to review progress just over one year later, another critic weighed in. Dmitri Ustinov, a grey party man if ever there was one, who had headed the armaments industry (and would later become the Soviet Union's defence minister), told Morozov: 'more than 3 years have passed, and the matter has hardly moved forward, and the tank has not actually been put into production and has not been handed over to military units.'

Early in 1967, Morozov wrote to Marshal Rotmistrov, begging him to keep faith in the T-64. The designer argued his principle that they needed to shun timidity in order to gain a great leap forward: 'he who has the courage to attack wins. Victory cannot be achieved in defence.' Since the T-34 had by that point become an icon of Soviet propaganda, Morozov exploited that in his arguments: 'The T-34 tank also had a difficult "birth" and everyone then advocated for the T-26 tank, which in terms of reliability was indeed better than the T-34 at that time. The war showed that supporters of the T-26 were wrong.'

Bit by bit, though, Morozov was losing the support of the military-industrial apparatchiks. Although they kept his project going, they also invested in alternatives, projects for tanks that, while similar in many respects to the T-64, had different powerplants: the T-72 and T-80.

Notwithstanding these developments, there was a consensus among the party bosses that Kharkov's costly project had been visionary in many respects. Realizing the scale of this achievement, they heaped Morozov with yet more honours: the Lenin Prize in 1967, and a second Hero of Socialist Labour medal seven years later. In his diary, the designer confessed his frustration with suppliers who had shown everything from laziness to downright obstruction, and with a hierarchy that made him responsible for the delivery of a revolutionary new weapon without giving him the powers he had enjoyed in the last war to drive the matter through. At the end of 1966, despite it being a turning-point year in the delivery of finished vehicles, he wrote: 'we work alone, there is no sense of comradeship . . . I don't have enough strength anymore!'

With the advent of the T-64, the Soviet Union stopped making heavies – what was the point, when it embodied superior firepower, protection and mobility to the T-10? Thereafter in the West too, production focused on a single design, which was termed a 'main battle tank'.

What, then, was the user reaction to this remarkable development? The Kharkov plant finally started releasing the T-64 to Soviet army units in 1966. Early consignments were given to battalions in the western parts of the USSR rather than in Germany, where the prying eyes of Western intelligence officers, and all manner of surveillance technology, would be focused on them. As we will see, the secrecy applied to the T-64 was extreme: it was a feature of intense Cold War rivalry and a feeling on the part of the Soviet general staff that they had developed something with a decisive qualitative edge.

At the time, a junior lieutenant named Vladimir Rezun was about to graduate from the Kharkov Guards Tank School. In an echo of the wartime system of training on the site of their factories, he and his class were held back to train for four months on the new vehicle. 'From the very first look we all liked the 125mm gun,' Rezun wrote later; 'upon closer acquaintance, our delight with T-64s had begun gradually to fade.' Once they started driving it, the breakdowns began. Rezun, who later defected to the West, writing books under the name Viktor Suvorov, continued: 'the engine itself was not only bad, it was disgusting. Several teams of workers and engineers, and a gang of designers, were sent along simply to maintain our one tank regiment.'

This direct support from the Kharkov factory was not in itself a revelation. After all, Chrysler had sent teams to Egypt in 1943 to help the British army with its Shermans. And in the later Cold War, Western weapons became so complex that such contractor assistance became routine. Morozov complained in his diary, though, that detachments of experts to military units just slowed the resolution of issues back at the plant.

In the Soviet Union of 1967, where robust engineering was prized, the difficulties encountered by the T-64 invited derision. Acknowledging the fragile nature of its engine, the Kharkov factory initially gave it a life of just 150 hours between major overhauls. Like many advanced systems with teething troubles, performance was improved and within a few years the powerplant was meeting a target of running for 500 hours without major failures, though even this was hardly impressive for what was meant to be the spear point of a Soviet blitzkrieg. Over time the temperamental new machine matured. Its engine was failing less often and it showed well in army reliability trials versus the T-72 and T-80. The users began to express more favourable opinions.

Another T-64 unit commander, Sergei Suvorov (not to be confused with the pseudonymous one), argued that much of the difficulty came about because Soviet crews lacked the know-how to operate such a modern weapon. Suvorov, who later wrote a book about the tank, argued that its development was a great leap forward: 'the adoption of the T-64 tank was comparable to the transition in aviation from piston engines to jet engines.' As for that engine, Suvorov argued, 'there were two main reasons for its failure in the army – overheating and dust wear. Both reasons occurred due to ignorance or neglect of operating rules.' Rigorous enforcement of maintenance procedures was needed. Mykola Salamakha, who served in a Soviet T-64 company during 1979–83, and later in the army of his native Ukraine, believed it was a highly manoeuvrable vehicle with many qualities. He too blamed 'poor knowledge and lack of skills' for the vehicle's teething problems.

Here we can see where Morozov's comparisons with the T-34 fell down. The wartime tank, particularly in its early versions, had plenty of flaws but ultimately gained the love of the Red Army because of its simplicity. It was soldier-proof, something you could hardly say for the T-64.

The steady growth in sophistication during the 1960s and 70s would pose challenges for all armies that relied on conscripts. From the IDF struggling with its new Centurions to the Soviet army and the T-64, inculcating recruits with limited training time proved to be a challenge. In Nato armies it was an important contributory factor in the transition to a volunteer, professional service.

While this long process of ironing out the T-64's flaws went on, the Kharkov plant's designers pressed on with another substantial innovation. The 125mm smoothbore packed a hefty punch, but beyond 2,000 metres its accuracy diminished markedly. Nato's rifled guns – the widely used 105mm and then the Chieftain's 120mm – could pick off targets several thousand metres away. Morozov therefore had to shift from his earlier stance eschewing long-range duels. Adding to the challenge, the Americans had fielded a guided missile that was fired from a 152mm tank barrel, a pricey innovation that they eventually abandoned, but one that, in the meantime, got them thinking in the Soviet Union.

Morozov's teams therefore delivered a dramatic breakthrough in 1976: *Kobra*, a guided anti-tank missile that could be fired through the 125mm barrel, just like a shell, but strike long-range targets much more accurately. The missiles and their advanced fire control system marked a significant jump ahead of the West. The US had developed a similar missile, but it couldn't be fired from the same gun barrel as high-velocity armour-piercing rounds.

Kharkov's latest version, called T-64B, did not go into large-scale production until 1977. It was a measure of the complexity of this technology, as well as the priorities of peacetime, that it took 20 years for the Kharkov designers, starting from their early experiments in the 1950s, to bring this concept to its ultimate fruition.

However, Morozov's rivals were not idling away this time. He was facing challenges in many quarters – not least from the other principal tank factories in Nizhny Tagil and Leningrad who resented the idea that Kharkov's T-64A would become the pattern for the whole Soviet army, the blueprint presented to them to be uncritically reproduced. Each plant had its own design teams, producing distinctive objekts or prototypes.

So, it was the Ural factory that gained favour with the Kremlin bureaucrats by developing the vehicle that became the T-72. It took many of the T-64's outstanding features – from a three-man crew to the 125mm gun and composite armour – but simplified its more exotic ones. The T-72 reverted to a V12 engine, and its suspension also followed a more conventional arrangement. The Leningrad plant got Marshal Rotmistrov's pet project, the gas turbine engine, mounting it in the new T-80.

Writing a few years before his death in 1979, Morozov bemoaned the way that the Soviet bureaucracy ended up producing 'three types of vehicles that are essentially identical in their combat and technical qualities, while creating exceptional complexity in production, operation, repair, training and supply of component parts'. However, because of the sheer size of their fleet, those running the Soviet defence industry could not overlook the need for production in multiple factories in order to replace obsolete vehicles – and the T-64 couldn't be delivered fast enough.

In time the Ural Wagon Factory in Nizhny Tagil would produce 2,500 T-72s a year, while Kharkov struggled to make more than 500 T-64s. Even so, it was the latter that remained the Soviet army's top-tier tank through the 1970s because of its advanced technologies.

The T-72 became an export success, being shared with Moscow's allies in the Warsaw Pact as well as Middle Eastern nations like Syria and Iraq. It was also seen in the Soviet military's shop window, the huge army it kept on Nato's doorstep, the Group of Soviet Forces in Germany. Meanwhile, the T-64 was not exported and for the first ten years of its service life was deployed only in western areas of the USSR among divisions that would have formed the second wave in any war with the West.

These precautions spoke to the secrecy with which Soviet generals wished to cloak the project, and the intensity of the arms race with its accompanying intelligence battle. Even after their forces invaded Afghanistan in 1979, they would not send any tank more modern than a T-62 there. And although Arab allies eventually got the T-72, the T-64 was vetoed: it could not be allowed to fall into Israeli hands. Because of this, it took many years for Western countries to understand the latest Soviet advances contained within the T-64.

At Potsdam in East Germany, the US, UK and France maintained military liaison missions throughout the decades of Cold War. The Soviet army had equivalents in West Germany. These outposts, each of which consisted of a few dozen staff, had been created in 1945 when the four Allied powers divided Germany into occupation zones. But quickly, as the atmosphere worsened, they had evolved into organizations tasked with sensitive missions.

Thus, the British Military Mission, known as BRIXMIS, was able to play a role in defusing tensions and in legalized espionage. Its vehicles toured about East Germany, photographing and observing Soviet or East German military movements, before returning to their villa, a relic of the time when Berlin's elite used Potsdam as a pleasant lakeside getaway. Observers from the military missions often departed from the agreed routes, and when they did, all manner of aggressive tactics were used to stop them.

The existence of Allied military zones in West Berlin, deep inside East Germany, also afforded plenty of opportunities for espionage. On the way in and out (by train, plane or automobile), there were lots of chances to gather images or electronic intercepts. It was a BRIXMIS observer who claimed the honour of spotting one of the first T-64As deployed to East Germany in 1976. Among the early images of this mysterious beast were a set evidently taken from one of the Western aircraft operating to and from Berlin.

Ten years after its first entry into Soviet army service it had therefore been sent to frontline assault units in Germany, with all the risks that held of the discovery of its secrets. Photographs could provide the basis for early analysis, but having the military missions in place meant there were opportunities to get right alongside, possibly even *inside*. The French, British and American touring officers prided themselves on such derring-do, operations often pursued at considerable risk.

Even two years after it appeared in East Germany, though, Western officials had a poor understanding of the T-64. A set of declassified intelligence assessments from this period allows us with hindsight to see the degree to which Soviet engineers had stolen a march on the West. One estimates the frontal armour of the T-64 at 150mm, whereas it was 200mm on the hull and double that on the turret. Another, from 1978, confesses the ignorance of British Defence Intelligence on this, noting: 'the difficulty was in obtaining an accurate vulnerability profile of Soviet tanks. The main concern is the degree of protection provided by the T-64/T-72/T-80.'

So, the technical team from Defence Intelligence back in London kept lobbying BRIXMIS for anything they could get. Recalling the pressure, Lieutenant Colonel Jerry Blake, who was serving in BRIXMIS at the time, said, 'Whitehall did not always appreciate that perhaps we were not in a position to rush on a T-64 to stick a gadget on it.'

In 1980, however, a pair of operators from the British mission were able to get right up to a T-64 that had been left overnight on a railway siding in East Germany. Pulling back a tarpaulin that covered it, they used a hacksaw to cut through the paintwork and into the hull armour. A specially designed tool was then used to drill into its surface to sample the metal. By this time Western suspicions that the T-64 used layered armour had firmed up, but they were still guessing at the details.

One area where they had greater certainty was in the matter of armament, specifically the development of long, fin-stabilized, armour-piercing shot, dubbed 'penetrators', for Soviet tanks. And what they had learned from their sensitive intelligence-gathering alarmed them.

Even the T-62's 115mm gun, firing a tungsten dart or penetrator, could overmatch the Chieftain's. A secret 1978 report noted the Soviet round would penetrate 330mm of armour at 1,000 metres, compared to the 295mm of the British gun. 'I do not believe we should allow ourselves to get over-alarmed about this development,' wrote one staff brigadier, emphasizing the various other aspects that play into a vehicle's firepower. But if that Soviet gun could do this, what about the bigger 125mm mounted on the T-64 and T-72? A secret British paper in 1978 estimated it could penetrate 540mm of armour, sufficient to defeat every Nato tank at the time from all angles.

Early in 1981, a daring BRIXMIS operation produced answers to many of the intelligence staff's uneasy questions about the T-64. Exploiting the May Day holiday, when much of the Group of Soviet Forces in Germany would be carousing drunkenly, they managed to get to a hangar on tank firing ranges near the East German town of Parchim.

Captain Hugh McLeod and Sergeant Tony Haw got out of their green 'mission car' and, seeing no Soviet sentries in sight, approached the building. 'None of us was under any illusions about how we would be treated if we were caught out of the vehicle,' McLeod would later reveal. Confrontations between these observing teams and the Soviet army had the potential to turn ugly. Aggressive driving, physical violence and gunshots had been used during previous dramas.

Rattling the hangar door, Sergeant Haw managed to defeat its lock, allowing them to get inside, where they found five T-64s. The young captain went to work, taking dozens of detailed photographs. As each film was finished he handed it to Haw, who in turn passed it out to their driver, who hid the precious rolls in different parts of their car. McLeod then turned his attention to the interior, using a tool specially made in anticipation of such a moment to unlock the commander's hatch. Dropping inside, he used a wide-angle lens to capture dozens of images of the interior.

The captain then noticed a training turret nearby, noting: 'I was able to take detailed photographs inside the simulator.' In some ways this was better than the tank itself because sections of its skin were cut away, offering better angles for taking photos. He also snapped numerous training posters on the wall. Once they were done, the two BRIXMIS operators returned to their car, getting away from the tank shed by driving along a disused railway line.

McLeod was decorated for this feat of daring. For the first time, Nato intelligence analysts had a detailed knowledge of the T-64's systems, its layout and the thickness of its armour. It confirmed the fears many had harboured. A couple of years later, an officer from the US mission, Major Arthur 'Nick' Nicholson, managed a similar feat with the T-64B, the later version of the tank that had the Kobra rocket launching system. But these contests to gather sensitive information in the face of determined attempts to stop them were to prove fatal.

In 1984, a French mission member was killed when an East German army heavy truck rammed his car at high speed as he drove along a road out of bounds to Western military liaison missions. And in March 1985, Major Nicholson, having dismounted to enter a Soviet training base in the hope of examining the newly deployed T-80, was shot dead by one of the sentries.

These sacrifices remind us of the deadly earnest in which the Cold War protagonists pursued their battle for advantage. Usually, it was a bloodless battle of intelligence collection, analysis, and the development of new technologies in response. But all of it, of course, was about jockeying for potential advantage if the Cold War turned hot.

One secret memo to Britain's Vice Chief of the Defence Staff in 1978, noting the evidence of how far the Soviets had got ahead in tank design, argued rather archly: 'it is obviously wise to design to "worst case" figures but we do not necessarily need the highest figures when explaining Soviet capabilities to our soldiers.'

As somebody who would have had to go to war in a Chieftain tank, had one broken out between the years of 1979 and 1983, I can attest that this conspiracy of silence was followed. We really were kept in ignorance of the degree to which the T-64 and T-72 could overmatch our own tanks in firepower and protection. We had complete faith in our thick armour, and the power of the Chieftain's 120mm gun. Ignorance is bliss, I suppose.

Whatever the teething troubles of the T-64's engine, we had plenty of issues of our own. The Chieftain's powerpack likewise took years to settle, even in 1979, for example, when I took part in an exercise under war conditions in Germany. Our squadron, required to 'road march' around 60 miles between two training areas, left seven of its 15 tanks broken down en route. My own, one of the survivors, could only travel at half speed due to a fault with its running gear. The British L60 engine, like that of the T-64, used a horizontally opposed piston arrangement cribbed from the pre-war German Junkers motor, only with six cylinders rather than five.

While we were all too aware of these issues of reliability, only those cleared at the highest levels understood how much better protected the T-64 was, or details about Britain's research in this area. At Chertsey, Britain's Military Vehicles and Engineering Establishment had for some years been conducting its own experiments on future tank design. By the 1970s its secret work on a laminar armour system, under the code name BURLINGTON, was stepped up a notch. Once made public, it was named Chobham armour after the road on which the base was sited.

What could be done to protect our tanks from the more powerful Soviet armour-piercing rounds? The same 1978 briefing note that advised keeping us users in the dark offered some hope: 'Director MVEE confirmed that some thickening plus a degree of BURLINGTON technology would permit such a performance.'

This research was to bear fruit with a composite armour package designed for fitting to the Chieftain. However, the solution was not delivered until 1986, some 20 years after deployment began of the T-64 with its laminar armour. The appearance of British-designed 'fin' or 'long rod penetrator' armour-piercing shot from the early 1980s would give our tanks the ability to penetrate 475mm of armour at 1,000 metres.

25. A Centurion Mark III in action in Korea in 1953, where its ability to soak up punishment while offering supporting fire won it the respect of infantry and tank crews alike.

26. A welder at work on a Centurion hull – the tank established a Cold War pattern, having several times the number of parts, and costing several times as much, as the Shermans bought just a few years before.

27. In its final incarnation in British army service, the Centurion AVRE (Armoured Vehicle, Royal Engineers) was used against Iraqi forces in 1991. It featured a 'bunker-busting' 165mm demolition gun and Israeli-developed bolt-on reactive armour.

28. Pictures taken during my visit to Jüterbog in East Germany in 1989, as the Soviet army prepared to withdraw, brought me face to face with T-64s for the first time.

29. The T-64's early service was blighted by issues of poor engine reliability. By the time I photographed this private at work in the compact engine bay, its performance was much improved.

30. Years later I spotted this T-64 at the Yavoriv training area in Ukraine. Much smaller and lighter than the Leopard 2s and Challengers they had received by this time, its crews valued its compactness.

31. The Merkava made its combat debut in Lebanon in 1982, and indeed it was there that the new vehicle was involved in its only tank-to-tank fighting to date.

32. By the time the Mark 3 version entered service in 1989 it was being reshaped to deal with irregular enemy forces mainly using shaped charge missiles, with a modular armour system that made the turret much larger.

33. The Merkava Mk 4 was the first tank to use an Active Protection System, called Trophy, to shoot down an incoming missile in combat. A radar for the system can be seen at the rear of the turret and above, to the right, a launcher for the projectiles it uses against threats to the tank.

34. The cost and sophistication of the Merkava Mark 4 is such that disabling such a vehicle is a propaganda prize for Israel's foes.

35 & 36. After years in which tank sceptics had once again predicted their imminent demise, the 1991 Desert Storm campaign provided an emphatic demonstration of the value of armoured warfare. Exploiting the technological superiority of the M1A1 the US military was able to impose vastly disproportionate casualties on Iraq's armoured divisions.

37. Repairing an Abrams in 2003. When facing Saddam Hussein's military the vehicle retained its superiority, but once pitted against insurgent groups some of its weaknesses were exposed.

38. Ukrainian workshops modified many of the vehicles sent by Western countries. Additional reactive armour bricks can be seen here, which critically attempt to protect the Abrams' thin turret roof armour from direct strikes by Russian drones.

39. Russian forces were quick to send Western armour captured in Ukraine back home as trophies. This M1 was an early loss.

During the late 1970s and early 80s, Britain shared the secrets of its armour research with its US and German allies. One of the Chertsey scientists remarks wryly that plans for an Anglo-German tank fell apart soon after their big reveal, with the Germans exploiting the new technology in the Leopard 2, which they exported very successfully.

Once its existence was revealed publicly in 1976, Chobham was extolled as a world-beating British invention. Roy Mason, then Defence Secretary, argued that it was 'a major breakthrough in tank defences . . . it should have quite an effect on the balance of land forces.' The *Birmingham Daily Post*, reporting this, added: 'there is no firm evidence that the Russians had developed anything like it.' Ignorance, again, was bliss. Quickly, members of parliament and journalists alike started asking questions about why the technology had been shared immediately with German and US allies or offered to export customers like Iran, as well as why it would take several years for it to reach the British army.

The Soviet army, though, was a decade ahead. By 1978, Nato intelligence estimated it had deployed 4,230 T-64s and 2,210 T-72s to its frontline forces arrayed against the Alliance. That same year the UK accelerated production of its first vehicle to be designed as standard with Chobham armour, the Challenger 1, entering service five years later.

Advocates of Chobham insist that it is superior to those earlier Soviet composite armours. Although aspects of this are still highly classified, one thing that has emerged is that the Chertsey scientists had discovered the value of thin air. Britain's composite built in gaps between a couple of its layers. The discovery here was that a penetrating plasma bolt from a shaped charge became 'unsettled' when it passed through the outer steel skin. If it then encountered air, this tiny wobble became amplified, bleeding energy off it before it struck the next layer of the laminate.

It was reckoned by intelligence analysts that when struck by a shaped or HEAT charge, the Soviet laminates of the late 1970s offered a 30 per cent improvement over homogeneous hardened steel. Chobham was somewhat better. But just as the West started to catch up or even exceed the performance of their armours, the Soviet scientists took another step in this game of tit-for-tat.

The technical intelligence battle was not one-sided. Soviet military intelligence was involved in a worldwide effort to snap up Western technology, with everything from tank night sights to guided missiles being acquired by intelligence officers in foreign countries, often sent back to the USSR by diplomatic bag.

In 1982, Israel invaded Lebanon, sending in a force of several divisions. Some of their vehicles used a new system of bolt-on armour panels designed to defeat shaped charges. They also used a new fin-stabilized armour-piercing round, fired from 105mm guns, against Syrian army tanks during fighting in the Bekaa Valley and on the outskirts of Beirut.

Soviet intelligence was able to obtain a specimen of the new Israeli tungsten armour-piercing shot, and it unnerved their experts. They would need to rethink their own protection levels. So, from 1982 onwards the Soviet Union searched for answers. In response to the new 105mm Western penetrators, they made plans to fit an additional layer of armour on the T-64's front. They also began testing a prototype system of 'active' or more accurately 'reactive' armour to counter new anti-tank missiles and other shaped charges. Early images suggested that a multitude of brick-like boxes had been mounted on these tanks. Western experts had a good idea what was going on, because it was very similar in appearance and function to the Israeli system fitted on their Centurions and older Pattons. Inevitably, the Allied military missions in Potsdam were told to report back, and ideally obtain a sample.

The first East German sightings came in from BRIXMIS in 1984. Detailed photographs showed a system of 256 boxes or tiles on each T-64B tank. Termed ERA or Explosive Reactive Armour, the principle was simple: when struck by a shell, notably one relying on the shaped charge effect, the brick exploded, and the effect of this blast was to disrupt the formation of the penetrative bolt made by that incoming round.

What the designers in Kharkov or Nizhny Tagil knew was that their vehicles were heavily protected, but only in the frontal aspect. A T-64 turret might be 400mm thick, or even more, around the gun, but its back was just 70mm, and in places the roof was less than that. Although they did not initially try to protect the rear with these kits, Soviet designers did want to make frontal armour even thicker, as well as covering the flanks and turret roof. Why? Because they knew during the early 1980s that Western countries were developing anti-tank weapons specifically designed for 'top attack'.

Sweden's Bill missile, for example, could be fired by a lone infantryman at a tank over a mile away. Close to the target, it climbed, exploding overhead (and with its shaped charge exploding at an angle) to pierce the roof. The US meanwhile was developing artillery missiles that would scatter bomblets over Warsaw Pact armoured formations, similarly striking their top armour with shaped charges.

This scientific contest evolving through multiple action-reaction stages – with laminar armour stimulating new anti-tank weapon developments, which in turn prompted the fitting of reactive armour – still had another stage to run. Seeing the deployment of Soviet reactive armour, Western scientists developed 'tandem' warheads, anti-tank missiles with a double shaped charge, the first to trigger the ERA, the second to penetrate the main armour underneath.

The T-64's reactive armour package weighed in at over 2 tonnes. Morozov's earlier weight limits were breached as the tank climbed over 40 tonnes. Inevitably there was an effect on its speed, but it remained 25 or more tonnes lighter than the Western tanks that appeared in the years after the Cold War ended.

Once Mikhail Gorbachev came to power in 1985, he set about diverting resources to the civilian economy as well as trying to reassure the West that he was not about to invade. Following the announcement of unilateral troop withdrawals from East Germany, I had the opportunity to go there as a journalist to witness the start of this process in 1989. There, at the tank sheds in Jüterbog, in a unit that had previously been earmarked to assault the British Army of the Rhine, I came face to face with an object we had studied for years, the T-64. Could I look inside? Be our guest.

It was so small. Inside, the commander's station felt almost as snug as being in a sports car. But it looked like a very efficient fighting machine. I was just pleased I had never faced one in earnest, and as the Berlin Wall came down later that year, knew I would never have to.

As the 1990s wore on, the T-64 fleet seemed to be cruising towards obsolescence without ever having fired a shot in anger. By the time production ended in 1987, 13,000 had been built. Because of the prohibition on exports, the Israelis never had a chance to capture one, and they were not encountered by the US-led Coalition during the 1991 Gulf War. The Afghan mujahideen ditto were denied the chance to salvage T-64 parts for the Western agencies in Peshawar that paid so generously for Soviet weapons' components.

The T-64's story might easily have ended as a sort of Cold War archetype. There were fighter planes or submarines that fell into this category too – weapons developed at a cost of billions that represented the acme of contemporary technological advancement but were destined never to be used. When the stock of available machines was divided between the republics as the Soviet Union collapsed, it was Russia and Ukraine that kept the largest T-64 numbers – and that meant that they would see action.

As the Russian onslaught of the 24th of February 2022 began, Ukraine's standing armed forces found themselves heavily outnumbered. To the north, east and south, dozens of armoured battle groups crossed the border on the Kremlin's orders, intent on rapid conquest.

One of the first actions took place near the town of Ripky, as Ukrainian forces tried to defend the north-eastern city of Chernihiv. Lieutenant Colonel Artem Linkov, in command of a T-64 from the 1st Tank Brigade, ordered his gunner to engage an armoured column to their front. 'For real?' the crewman asked. 'Yes, for real,' Linkov replied, 'it's not training. So he fired and the BMP [or infantry fighting vehicle] turret blew into the air.'

They had been committed to action after days of contradictory orders. Despite Western warnings, President Volodymyr Zelenskyy had been reluctant to order mobilization, believing President Putin was trying to intimidate them.

While the 1st Tank Brigade had a strength on paper of around 100 tanks, only a couple of dozen runners were available. Leading three of them forward, Linkov reflected, 'OK, well maybe I will survive.' Two of the T-64s broke down before they could open fire. Although modernized, these vehicles were 40 years old.

The Russian column at Ripky was heavily engaged with artillery, losing dozens of vehicles. Elsewhere on the Chernihiv defence line, at the Pivtsy airfield, a group of five T-64s reinforced the defenders, fighting for days under heavy Russian shell and tank fire.

To the far south, meanwhile, another Ukrainian tank man, Lieutenant Yevhen Palchenko, was earning himself the Hero of Ukraine medal. The 23-year-old officer, commanding a company of T-64s, had to lead a larger group of troops out of a Russian encirclement. On the 24th of February, fighting with the key bridge over the Dnipro at Kherson to their backs, Palchenko's people realized they were about to be encircled. During the withdrawal to positions close to the crossing, five of his eight tanks ran out of fuel. In this sector, Russian forces did far better than to the north, advancing rapidly and causing chaos in the Ukrainian command. Realizing the net was closing, Palchenko broke out of the encirclement with a force that was eventually reduced to two tanks and accompanying infantry.

The many engagements that took place during the fighting of 2022 provided a final endorsement of Morozov's ideas. The T-64, brought into service decades before, remained a viable fighting vehicle. And it's not as if the Ukrainians could have chosen another – at the time of Putin's invasion 720 out of the 850 tanks in

Ukrainian service were T-64s modified to some degree over succeeding years.

As Morozov himself had opined, the fact that this vehicle, the T-72 and the T-80 all had the same main armament, automatic loading and the same levels of armoured protection meant there wasn't a great deal to choose between them. The Ukrainian army owed its successes at the gates of Kyiv or Chernihiv to better training, the use of initiative, as well as superior combined arms tactics, and in particular the exploitation of drones to find targets for artillery fire missions.

The 2022 fighting in Ukraine would also remind everyone that modern inter-state warfare would consume enormous numbers of lives and weapons. Russia's tank industry had by 2021 fallen to a production rate of 200 vehicles a year. In Europe, the figure was far lower. And while the Morozov legacy lived on, in terms of believing that armies needed a large number of compact, mobile tanks packing a big punch, in other countries the tendency towards ever smaller numbers of heavy, highly sophisticated and therefore extremely expensive vehicles had continued. And it was in the Middle East, seen by many as one of the most suitable environments for armoured warfare, that these trends reached their height.

Merkava

Entered service: 1979
Number produced: 2,200
Weight: 60 tonnes (early variants)
Crew: 4
Main armament: 105mm gun; 120mm on later marks
Cost: $1.8m in 1990; the latest Mark 4 version $10m in the early 2020s

Merkava

At around 2 p.m., Colonel Hagai Cohen got his orders. He had dozens of tanks – in fact, among the mixed fleet in his brigade, he had 15 of the new Merkavas, pride of the Armoured Corps. However, hastily ordered up by his superiors, he didn't have any infantry under command and would only get fleeting artillery support. Despite that, the colonel had been told to take the Lebanese town of Jezzine as part of a much larger Israeli operation.

His armour rumbled towards the Lebanese resort town from the west, seeing it laid out in a deeply incised valley beneath them. To the east and south of Jezzine were the steep slopes, rocky cliff in places, of the Tumat Niha hills. With movement channelled, and the built-up area overlooked by many lofty vantage points, it was dangerous country for an all-tank brigade.

Two days before, on the 6th of June 1982, Israel had invaded Lebanon. Their aim was to drive back Palestinian armed groups with whom they had traded fire for years. But they were also about to confront a 25,000-strong Syrian army force that occupied parts of the country. Moving north, the Israelis initially deployed their armoured divisions, a force that would eventually total more than 70,000 troops and 800 tanks. In many ways this mechanized juggernaut was unsuited to the hill country that it had to cover in order to reach Beirut.

Major General Avigdor 'Yanush' Ben-Gal, one of the higher commanders, described the advance as 'characteristic mountain fighting, in and out of many narrow byroads and pathways. We had to proceed slowly, at infantry walking pace.' It was hardly the ideal way to commit the IDF's new tank.

Now, despite his inability to mount an all-arms operation, Cohen had been ordered with minimal notice, few maps and no proper reconnaissance, to take the town with his 460th Armoured Brigade. Summoning his battalion and company commanders, it was explained that there were only a few hours of daylight left, and their priority was to take the town quickly since it was a gateway to the Bekaa Valley, the Syrian army stronghold to the north-east.

Cohen's brigade was an unusual one. It was mobilized by drawing together battalions from Armoured Corps training schools, elements that prepare soldiers for service in different fighting vehicles. His hasty battle plan involved the 198th Battalion leading the way into Jezzine. This unit, normally the school for vehicle commanders (i.e. sergeants), had a mixture of Merkava and Centurion platoons (the Centurion a modified version called a *Sho't* or whip in Hebrew). It would be supported, as it dropped down into the town from the high ground west of Jezzine, by the 196th Battalion. In peacetime this was the school for tank platoon commanders, young officers, having a mixture of Merkava, Centurion and Patton tank platoons.

During the following hours, therefore, they would receive lessons in the merits or failings of these different machines, as well as the challenges of taking on such a complex task with so little preparation. Captain Gidi Sturlesi, a company commander in the 196th, noting that its leaders were instructors and their 'crewmen' officer cadets, believed 'it was without doubt a quality battalion, the most elite unit of the armoured corps'.

As befitting the ethos they sought to inculcate in their pupils, officers led the way. The first company of the 198th rumbled down into the Lebanese town; Colonel Cohen, with his command element of three Merkavas, drove just behind the lead platoon. Very quickly, as it entered the outskirts, this column came under fire from surrounding hills and buildings.

Syria had garrisoned Jezzine with an infantry battalion. Several companies of commandos, armed with anti-tank missiles and rocket-propelled grenades (RPGs), had also been sent in. Early in the morning of the 8th, these had been reinforced with a battalion of T-62s, around 30, of the 51st Armoured Brigade that moved out of the Bekaa to secure this important hub. The fight that took place that afternoon would therefore pit the brand-new Merkava against a modern foe for the first time.

Although the Israelis knew there were Syrian troops in the town, they had not been told anything about tanks, or indeed the commandos who carried numerous anti-tank weapons. Despite some skirmishes, major fighting between the IDF and Syrians had not yet begun.

As the first company of the 198th Battalion rumbled into Jezzine, it triggered that battle. The radio came alive with contact reports, and the valleys echoed with the firing of tank guns. Colonel Cohen found himself in the midst of it: 'I was very forward, just behind the first platoon. The firing distances were very short . . . and our knowledge of the enemy very rough.'

When the two follow-on companies of the 198th tried to head in, they were met with a hail of RPGs and small arms fire. Many of the Israeli crews were out of their hatches, desperately trying to keep situational awareness. Syrian troops aimed at them, and in some cases lobbed grenades from upstairs windows. After suffering several hits, these two tank companies pulled back. There was no infantry with them to clear the buildings. Cohen radioed the 196th Battalion, telling them to bypass Jezzine if they could.

Out to the west, moving forward on the high ground, Captain Sturlesi stopped, then dismounted from his tank to confer with one of the 198th's tank commanders who'd just pulled back. He attested to the fury of the Syrian defence and narrowness of the streets. When Sturlesi resumed his advance, he was therefore looking for a right or southwards turn that would allow him to bypass the centre, while staying above it on the valley side. Moving forwards, his company soon came under highly effective fire.

Sturlesi's tank and those of most in his company were M60 Pattons (*Magach Gal* in IDF terminology), modern enough, and he reckoned with a rather better fire control system than the new Merkava. But that didn't save them from what followed. 'One shell landed just in front of me,' Sturlesi remembers. 'We just thought it was mortars or RPGs.'

Just after ordering his tanks to deploy smoke grenades to conceal themselves, his Patton was struck, flakes of steel from the impact flying into his face and upper body. As he dismounted to take command of another tank, another explosion echoed across the hills, this shot striking his company second-in-command's tank with disastrous effect. Both the commander and his gunner were killed. A further shot from above, and another Israeli tank was hit. Sturlesi's company had lost three vehicles by the time it was able to back out of the danger. The captain only realized later that, exposed as they drove along a hillside road, they had been ambushed by T-62s on the heights to their south.

Here we should recall that, although the Syrians had an even more modern vehicle, the T-72, the T-62 was still highly potent, particularly in respect of its firepower. It was the T-62's 115mm gun after all that had caused such concern in Nato intelligence circles just a few years earlier: firing a 5.39kg fin-stabilized armour-piercing penetrator that left the muzzle at 1,680 metres per second, it could penetrate more than 300mm of steel. That was sufficient to dispatch any of the tanks the Israelis had taken into Jezzine that day, including the Merkava.

That Syrian armour had initially deployed on the high ground, facing southwards, towards Israel. However, realizing that the Israelis had turned their flank and were entering Jezzine from the west, and a few miles to their north, they turned their T-62s around, discovering that their vantage point atop the massif allowed them to engage Sturlesi's tanks at ranges of 1,500–2,000m, which was ideal from their perspective.

When he later examined his own knocked-out Patton, Captain Sturlesi discovered the finned penetrator lodged in the turret side, just above the junction with the hull. He could not explain why it hadn't gone fully through and killed the whole crew, putting it down to luck.

At this point in the late afternoon, though, having extricated the remnants of his company, Sturlesi followed on behind that of another captain, Tzur Maor. He had nine Merkavas and a platoon of (three) Centurions under command. For Colonel Cohen, pinned down, isolated on the eastern side of the town centre, jeopardy was increasing: 'it could have got very serious.' If the Syrian commandos were able to press their advantage, they might get wiped out.

Captain Maor, though, had at last succeeded in finding a route that would take them south, towards the higher ground from where the T-62s could dominate Jezzine. Like many a hill road, it zig-zagged as it climbed up through the outskirts of Jezzine between the olive groves beyond. There followed a series of engagements, many at very close range, as the Merkava rounded corners, encountering T-62s on the road ahead.

It was now that one of those issues of tank design that seem so dry and technical on the page started to make the difference between life and death. Because of the engineers' desire to give it the lowest possible silhouette, the T-62's gun cannot depress by more than six degrees. In order to hit the Israeli tanks climbing the steep slope below, the Syrian crews had to drive forward onto the slope, thereby exposing the whole of their vehicle to enemy fire.

Captain Sturlesi, following up, watched Maor up ahead climbing: 'he met a few tanks on the road itself, he was always first to fire . . . he did an unbelievable job.' Burning T-62s lined the road. Trying to pass one, Maor's huge tank dropped off the roadside, shedding a track. The officer jumped down and sprinted to another Merkava, taking it over and pushing the last couple of hundred metres to the summit, having already destroyed six Syrian tanks. As his own crested the rise, a T-62 struck from only 200–300 metres away: 'no tank could have withstood this, I saw it jerk back as the Syrian shot hit it,' recalls Sturlesi. Maor was killed, along with two other members of his crew, only the gunner surviving. However, he had successfully led the capture of the heights.

Moments later, a further misfortune befell them, a spectacular own goal. Hearing of the tanks' difficulties in Jezzine, a higher commander had requested an airstrike. But there was no forward controller with the column climbing the hill, so when the jets roared overhead, attacking Syrian and IDF vehicles alike, the road became blocked and the group climbing the southern massif was split.

The seven surviving Patton tanks from Sturlesi's and another company pushed on past the wreck of Maor's tank to a junction beyond the crest. During the night these crews held their position, repelling several attacks by Syrian commandos. They felt fortunate at least to have had effective night vision sights in their American-built vehicles because the Merkava Mark 1's lack of such optics would have been keenly felt during this action. The following morning the Syrians abandoned the town and its surrounding heights.

Estimates of tanks knocked out during the battle were ten Israeli and double that number of Syrian. Many of the IDF vehicles were recovered, but there were four permanent losses, including two Merkavas. The Israelis took satisfaction that their new vehicles had survived multiple RPG strikes, though one had been lost in the street fighting and another, Maor's, to a T-62 firing penetrating shot.

So, Israel's new war machine emerged from its baptism of fire. Nobody knew it then, but Jezzine in 1982 was likely the last time the IDF fought a tank-to-tank battle, and probably the only time that a Merkava has been knocked out by a kinetic energy, penetrating shot as opposed to various shaped charges, anti-tank missiles and roadside bombs.

Jezzine had been secured, and battle joined with the Syrian army. In terms of military history, the main event was just beginning: a major fight between the Israeli air force and multiple batteries of Syrian surface-to-air missiles in the Bekaa Valley, as well as dozens of fighter jets. The neutralization of these defences by methods that included drones and anti-radar missiles was a significant military achievement. However, the conflict is largely remembered today for its disastrous impact on the people of Lebanon, the

Israeli occupation of the country's south that followed, and the emergence of a radical resistance movement among the Shia population, the Party of God or Hezbollah.

During the invasion of 1982, the Israelis lost several dozen tanks, but managed to defeat 400 Syrian ones. As so often in armoured warfare, the issue of who held the field had a big effect on this total, since many lightly damaged or abandoned Syrian machines fell into Israeli hands.

These engagements were to prove a pivot point in the development of tanks. In June 1982, the Middle East was still a prime theatre in the wider Cold War technological contest. The Syrians were delighted to pass on to their Soviet advisers examples of the latest Israeli 105mm armour-piercing shells, or the new armour kits fitted to their tanks. For their part, the Israelis shared their knowledge with the US about defeating the latest Soviet-made air defence systems, and it was much to their chagrin that they were unable to recover any of the T-72 tanks knocked out in the Bekaa Valley. While it took a while longer for the Cold War to end, Lebanon would be the place where questions about what tanks were for, and how that should inform their design, would start to re-emerge.

What the leaders of the IDF were sure about at the time, though, was that the Merkava tank's performance against the T-62 and lesser enemies during the 1982 fighting was impressive, justifying the national investment in this technology. They would prove particularly pleased with the way this home-made tank had protected its crews from so many hits. If they had lost a few Merkavas, well, the hurried way they'd committed them in Jezzine meant their tactical approach hadn't been the best. And nobody expected it to be invulnerable.

So why put this tank in the pantheon of great designs? Firstly, because it represents something very important in the late industrial age: a war machine designed above all to protect its crews and thus reduce casualties. Secondly, the Merkava that rolled into Lebanon bears little resemblance to the one that's been used more recently in Gaza or Lebanon, and in this sense its evolution gives us clues about the future of armoured vehicles.

This story of constant change saw the original ideas of its purpose evolve rapidly, along with the technology it used. Indeed, in a time when many countries have asked whether it's still worth having tanks, the Merkava came to show the way in defining answers to that dilemma.

The ability to adapt so quickly rested in part on the fact that the Merkava was sovereign technology. That had happened because back in the late 1960s the Israelis found themselves blocked from buying the foreign tanks they wanted.

Following the cancellation of their order for Chieftain tanks late in 1969, as we read in the Centurion chapter, the Israelis moved swiftly to find an alternative solution. A working group under Major General Tal was tasked to explore whether Israel was even capable of manufacturing its own tank, and if it would be affordable to do so.

By the late summer of 1970 Tal had persuaded the government that the answer to both questions was yes. Israel lacked locomotive factories like the ones in Kharkov or Kassel that had turned their heavy plant to the making of war machines. Huge automobile lines or steelworks on the US pattern were lacking too. But the Israeli military had been engaged in increasingly ambitious rebuilding of Centurions, Shermans and Pattons at the Tel Hashomer plant near Tel Aviv. This had also given them plenty of practice in heavy engineering and the integration of complex systems.

As for the management of this new endeavour, the army itself would do it. The Ordnance Corps would act as the prime contractor – an arrangement that might have produced disastrous cost increases in many Western countries, but which Tal reckoned would help keep the costs down. The socialist ethos of the kibbutzim was still strong in early 1970s Israel, and the idea of the government running such complex projects did not strike them as unrealistic. It could serve to contain cost by eliminating the contractor's profit margin.

In later years the tank project would come to employ more than 6,500 people directly, drawing components from 220 subcontractors. Given that the country had fewer than 4 million citizens at the time, it was a major national endeavour.

Like the inventors at Fosters or Renault decades before, in the spring of 1971 Tal's design team built a wooden mock-up of their vision. It survives to this day. Israel's tank was designed from the outset to put crew survival first. Tal believed that 'we can provide protection to people inside a tank many, many times more than to the entire system we call a tank.' Lieutenant Colonel David Eshel explained: 'every part of the tank – be it fuel, ammunition bins, tool boxes or other equipment, would play its part in providing all-round protection for the crew.'

In Tal's view, this represented a rejection of old ideas of balance between firepower, protection and mobility: 'this is our original approach, the most important difference between our design and all the rest in the world.' While the general turned project manager – and one-time student of Talmudic texts – could occasionally turn theological about this, Tal's friend, the American strategist Edward Luttwak, put it rather concisely: 'he wanted a different tank, not a balanced tank, balance between protection, firepower and mobility, but unbalanced, really protected.'

The surviving mock-up reveals that this philosophy guided them early on to a conclusion that broke with design practice since the days of the Renault FT: that the use of a rear engine compartment should be rejected in favour of one at the front. Locating it and the transmission there not only meant greater protection for the people behind these big bodies of metal, but you could have a hatch at the back that would ease the loading of ammunition or escape of its crew. The thinking underpinning the design would be tested when, at an early stage in development, Israel was attacked by Egypt and Syria.

The 1973 October War pitted huge armoured forces against each other, including a total of something like 6,000 tanks. General Tal would later claim, given the proximity of the conflict's two fronts, that a greater number of armoured vehicles had been employed in this area than across the Kursk front in 1943, making this the greatest tank battle in history.

In fact, the geographical differences meant that the fights on Israel's northern and southern fronts had very different characters. Surprised on the holiest day of the Jewish calendar, the Day of Atonement or Yom Kippur, Israeli forces on the Golan Heights, an upland plateau strewn with huge volcanic boulders, had used fighting positions prepared long before that offered their armour numerous 'hull down' positions.

On the Suez Canal, meanwhile, a successful Egyptian assault, forcing the water barrier which averages around 200 metres in breadth, quickly overwhelmed a line of infantry forts, to which the Israelis responded with local tank attacks across open desert. On both fronts there were dramatic initial setbacks, but the Israelis managed to hold on for the days that their reserve divisions needed to mobilize, and the tables were then turned.

In the Sinai, the Egyptians received initial armoured counterattacks with a barrage of anti-tank missiles. The AT-3 Sagger (as Nato called it) was carried by its operator in a 'suitcase', the lid of which provided a base for the missile, which was steered to its target by means of a joystick, the commands reaching the rocket via a fine wire that unspooled as it flew up to 3,000 metres from the firer.

The Egyptian army operation had been prepared very thoroughly. Having forced the water obstacle, infantry teams went forward around half a mile to the sand berms that ran parallel with it, barriers prepared by the Israelis as cover for their tanks. These soldiers carried RPGs and some Saggers, but many others were kept on the western or Egyptian side of the canal, able to provide overwatch for the units that had crossed.

The Israeli 14th Armoured Brigade, equipped with 91 Pattons, was tasked to provide support for infantry forts along the length of the front. Moving forward in platoon and company groups, they soon came under a withering fire. By the end of that first day just 14 of the brigade's tanks were still working. One battalion, racing forwards, lost 22 of its 25 Pattons to Egyptian rockets and missiles in a few hours, leaving the desert dotted with blazing wrecks. During the first 48 hours of the war, the standing IDF armoured division in Sinai lost 200 out of its 300 tanks.

The success of Egypt's anti-tank missile teams would cause much collective soul-searching in Nato defence ministries, leading to fresh pronouncements that the tank was dead. But the Israelis had temporarily lost sight of what any veteran of the 8th Army from 30 years before could have told them, that trying to rush anti-tank defences in this way was suicidal, so much so that the British crews nicknamed them 'Balaklava charges'.

Blame for this was placed on none other than Talik, Major General Tal, whose idea of what he called 'the totality of the tank' had swayed IDF doctrine in the years after the Six Day War. The army became highly mechanized, infantry was put in a subordinate role, and Tal's belief that, in deserts in particular, 'armour shock' would win the day permeated the organization. In 1973, Israeli historian Abraham Rabinovich noted, 'The concept of armour shock had been turned on its head by the Egyptians' new weapons and tactics – it was armour that was being shocked by infantry.' Like some of the British theorists before him, Tal had come to overestimate the power of the tank.

Once the Israelis adopted proper combined arms tactics, raking the dunes with artillery and suppressive machine gun fire, the Egyptians' task became much harder. Sagger operators were cut down or got in cover, and were unable to 'pilot' their missiles towards targets. After the IDF regained its balance, it mounted a large-scale counter-attack, crossing the Suez Canal in the other direction, trapping the Egyptian 3rd Army and bringing their foe to the negotiating table.

In the north, the battle was arguably even more desperate during the first 48 hours, but the outcome more instructive for the future tank that Tal was developing. During the initial assault, two

Israeli armoured brigades, operating 181 upgraded Centurions and Shermans, had been assailed by three Syrian divisions with a total of 1,400 T-54/5s and newer T-62s. Although there were also some defensive forts garrisoned by infantry on the Golan, the early fight was largely a tank battle. One of the most vivid accounts of it was left by (then) Lieutenant Colonel Avigdor Kahalani, a battalion commander in the 7th Armoured Brigade.

During the first 36 hours of the battle, Kahalani had seen his force ground down by successful enemy combined arms tactics. 'The commanders were tired and tense,' he would write later, 'ammunition stocks were dwindling. Officers were dying, and each and every one of us was now fighting for his life.' His superior, in charge of 7th Armoured Brigade, admitted by this point that control had largely broken down: 'Our tanks, even if they held on, would fight as individuals.'

By day three, there were just seven working Centurions in Kahalani's battalion. Gathering a couple of other vehicles that were moving up from being repaired, he took on a Syrian armoured brigade, holding a key hill in what was known as the Valley of Tears. His surviving Centurions succeeded in turning the tide, destroying dozens of armoured vehicles. By the 10th of October his brigade (under the same Yanush Ben-Gal who we glimpsed at the start of this chapter) had just 35 of its 105 tanks still operating.

The battle fought by the 188th Armoured Brigade on the southern part of the uplands was even more desperate. Starting the war with just 76 tanks because of maintenance work, it was overrun during the initial Syrian assault. One of its officers, Major Shmuel Askarov, leading three Centurions up to firing positions on the war's first afternoon, noticed that the other commanders wouldn't follow him. Running back from his own vehicle and mounting one of the turrets, he put a gun to the commander's head, ordering him forward.

It was claimed that Askarov's tank destroyed 35 Syrian ones in the two hours that followed. But the Syrians managed to flank his positions, knocking out Askarov's Centurion and flowing past the other surviving vehicles. Soon, leading Arab elements had crested the Golan Heights and were looking down on the Sea of Galilee. For some hours the Syrian aim of reconquering land lost in 1967 seemed attainable. In this area Lieutenant Zvi Greengold was sent up from the Galilee with four Centurions to prevent this breakthrough. In fierce close-range fighting his platoon destroyed dozens of Syrian tanks. Greengold and Kahalani were later decorated with their nation's highest award for gallantry.

By gaining surprise, the Arab attack had caused a brief existential crisis. Major General Tal's verdict was that 'Israel's armour was forced to contain the enemy offensive by piecemeal counterattacks against large, well organized forces – in direct contradiction to the principles of armoured warfare.' This not only meant that traditional ideas about using armour en masse had been jettisoned in those desperate early fights, but that the IDF's tank arm had suffered greatly disproportionate casualties as a result.

Israeli accounts give varying estimates for vehicles lost. In part this is because of the recurring question of whether a machine was irreparably destroyed or could be repaired, but one of the higher estimates is of 1,063 tanks knocked out, 472 of which were permanently written off. As for the human cost, over 2,600 IDF troops were killed in the Yom Kippur War, with something around 830 being 'tankists', who suffered double that number of injured. The salient point, though, given the wide range of different units involved, is that the Armoured Corps had been hit hard – reflecting the period, particularly on the Golan, when its crews stood between the Syrians and victory.

The blood price paid in 1973 helped to seal the public case for developing a new Israeli tank and would have a great influence on its design. At times the IDF had felt they were up against a technologically superior enemy, the Soviet-made vehicles having night vision equipment, and the T-62 in particular a better gun. Colonel Benny Michelsohn, the Israeli Armoured Corps historian and a company commander in those 1973 Golan battles, argues that this compounded frustration with allies who would not provide the most up-to-date equipment; furthermore, 'it was clear that no improvisations or rejuvenation to tanks of the 1940s and 1950s would suffice.' The die was cast in terms of building an indigenous tank. This would take the Hebrew name for a chariot, the *Merkava*.

Given that the basic configuration for this new vehicle had already been set, the October War experience involved changes of detail rather than a full-scale return to the drawing board. During 1972, the Ordnance Corps had assembled its team of subcontractors and in May 1973, having chosen the same American-made 900hp diesel engine used by the IDF's newer Pattons, some had been bought for the prototype. The gun too, an Israeli-made version of the British 105mm L7, also gave commonality with other vehicles in the fleet – though the Merkava was designed to accommodate something bigger in the longer term.

Principal changes made as a result of the recent conflict included rethinking armoured protection and developing different tracks. After studying hundreds of knocked-out vehicles, Tal made some changes to the design's internal arrangements, notably to make better use of stowed items to enhance crew safety. On the Golan, the Israelis had found that tracks with rubber pads (of a kind fitted to American-made tanks) were not tough enough to survive on terrain strewn with basalt rocks. The Merkava would therefore have all-metal ones and running gear built to take the unique stresses of that environment.

Experience of fighting the Syrians had also reinforced the value of having a hatch at the rear of the hull. Some Centurions had shot their way through three loads of ammunition, pulling out of their fire positions to reload each time. Since the new vehicle could be reloaded through the hatch at the back of its hull, crewmen avoided being exposed to fire when passing shells up to turret hatches as they had to do with other tanks. By putting the 105mm rounds in four-shell cells it could be done faster too. The Mark 1 tank could carry 62 main armament shells, a hefty total influenced also by 1973 war experience. Leaving some or all of this hull ammunition space empty, other roles were opened up, for example for the evacuation of crews from knocked-out vehicles.

Moving swiftly after the conflict, the decision to press ahead with the project was made in May 1974, and the first prototype entered testing seven months later. But the sourcing of parts and integration of such a complex vehicle took time. It was nearly five years before the first production vehicles entered service in 1979. Colonel Michelsohn argues: 'but for General Tal it wouldn't have happened. He had a vast knowledge and a vision that nobody else could match.'

The Merkava therefore beat the M1 Abrams into service by a matter of several months, which is why this chapter precedes the one on the American vehicle. In many ways, though, they are parallel stories in which two advanced nations, sensitive to their casualties, made the protection of those operating these war machines their highest priority.

So what about the points of difference? The American vehicle embodied superior electronics, for example a thermal imaging gunsight, when it appeared. It had proper composite armour too, secret technology that had not yet been shared with Israel. The Mark 1 version of the Israeli vehicle used spaced plate, creating a gap between the outer and main layers of armour. Stowed materials from tools to fuel were put into this to enhance the protective effect. But the Merkava was designed to be more rapidly adaptable, and in the aftermath of the 1982 Lebanon War the rethink started immediately.

The Merkava Mark 1 had performed perfectly well in battle, but it had hardly been an unstoppable juggernaut, impervious to the fire of its enemies. Israel remains secretive about its operational losses, even those of 40 years ago. But open-source analysts suggest seven Merkavas were knocked out during the 1982 fighting, at least one by a Syrian T-62 and others by Sagger missiles or rocket-propelled grenades fired at vulnerable spots.

However, the Israelis have revealed that they were satisfied that their home-made 'chariot' proved harder to kill than the Patton and Centurion tanks making up the bulk of the fleet operating in that conflict. IDF operational analysis suggested 48 per cent of the Pattons hit by enemy anti-tank fire had been knocked out, 38 per cent of Centurions, and just 20 per cent of Merkavas.

The 1982 war led the IDF all the way to Beirut and produced a whole set of geopolitical consequences, including the evacuation of the Palestine Liberation Organization from that city, a lengthy occupation of southern Lebanon and a good deal of international opprobrium for Israel's actions in general. This being a history of the tank and its place in conflict, we will focus on the extent to which the war was a turning point in that story.

Although the forlorn Israeli charges of October 1973 were regarded by many as the seminal moment in that long contest between the shaped charge and armour, it was in Lebanon in 1982, arguably, that this method of attack demonstrated even greater power. Anti-tank missiles fired from helicopters had been used for the first time. And in another baptism of fire, a column, reportedly of nine Syrian T-72s, was knocked out in the Bekaa Valley. This miserable first showing by new-generation Soviet armour was claimed by some Israeli generals at the time to be the work of Israeli tanks and even the Merkava. However, later analysis strongly suggests that the Syrian column had been engaged by the 409th Anti-Tank Brigade, a jeep-mounted paratrooper element equipped with anti-tank missiles, and some similarly armed helicopters.

While the question of whether the latest 105mm armour-piercing shot could penetrate laminar Soviet armour therefore remained unanswered, it had been emphatically shown that the US TOW missile (it stands for Tube launched, Optically tracked, Wire guided), with its large High Explosive Anti-Tank warhead, could do so. As far as Israel's fights with neighbouring states were concerned, as we have seen, Jezzine in 1982 marked the last engagement (at least at the time of writing) between tanks. Instead, the Syrian army avoided confronting the Israelis during its subsequent actions in Lebanon, for example in 2006, while the emergence of a powerful Shia militia, Hezbollah, saw the IDF facing new threats in the south. Backed by Syrian intelligence and the Iranian Revolutionary Guard, this force would become highly proficient in guerrilla and small unit tactics, including against armour.

Initially, though, in the wake of its 1982 operation, the Israelis built an improved Mark 2 Merkava that rectified some of the earlier model's protective weaknesses. There was thicker armour at the rear, and an ingenious system of hanging chains on the back of the turret was employed to stop RPG gunners exploiting that 'shot trap' between it and the hull.

In the years following the 1982 war, the threat evolved, as did the requirements of the IDF. Its armoured corps had swollen up to close to 5,000 tanks after the Yom Kippur War, and despite the small size of its population, a fully mobilized IDF could field ten armoured divisions, compared to Britain's four. During the 1980s the pendulum began to swing back the other way. Egypt, the neighbour with the largest field army, had made peace with Israel in 1979. Thousands of vehicles that had been kept going by retrofitting new engines or armament, from Sherman to Centurion tanks or M3 halftracks, were no longer fit for purpose and would cost a fortune to replace. During the wars of 1973 and 1982 many commanders felt the balance between armour and infantry needed to be redefined, as would ideas about protecting them.

The realization that infantry had to be protected to the same degree as tank crews led the Israelis to evolve vehicles quite different to those in Western armies, and in many cases to cannibalize old tanks to do it. Cold War armoured personnel carriers (or APCs) often had ludicrously thin armour compared to tanks; for example, 44mm of frontal plate on the US M113 or just 10mm on the Soviet BTR-80 (versus 250mm on the M60A1 turret front or 400mm on the T-64).

Given that more lives were at stake in these machines, this struck the Israeli military as being quite illogical. A few years after their occupation of southern Lebanon began, they started issuing the *Achzarit*, a T-55 tank with its turret removed, and the weight more than compensated for by piling on great thicknesses of armour. Later, Tal pushed ahead with the *Namer* project, a heavy APC based on the Merkava hull. To give some idea of how different its protection levels were to Western vehicles, it weighed in at 63 tonnes, compared to the M113's 13 tonnes. Namer is thickly armoured in all aspects – so much so, that the access to the rear hatch between two thick banks of protection looks almost like a corridor.

These vehicles had no equivalent in Western armies or indeed the Russian one. Most countries that were serious about armoured warfare chose to develop 'mechanized infantry combat vehicles'. They were designed to give these soldiers a better fighting chance, usually with a turret mounting a gun and anti-tank missile system as well as improved armour – albeit far less thick than that used on those IDF machines.

Operations in occupied southern Lebanon had underlined the need to plan for very different kinds of battles. There (and later Gaza), this was about an unseen enemy that might attack from any range or point of the compass using shaped charge weapons capable of piercing even the thickest armour. But if a major war broke out against Arab states that were re-equipping with very modern foreign-made fighting vehicles, the Merkava would be pitted against a more traditional type of foe: an enemy to the front where victory in the tank-to-tank battle would depend on who had the better high-velocity armour-piercing rounds or fire control systems.

The attempt to square this circle resulted in the Mark 3 design, embracing so many changes that it was almost a different vehicle. It entered service in 1990, and it marked to a considerable extent the technological catching up of Israel with the US, which fielded the M1A1 version of its Abrams at almost the same time. Both of these new variants used the 120mm smoothbore gun originally designed by the Germans. Both used modern fire control and night vision equipment too. By adopting a 1,200hp engine, the Israelis had closed some of the gap with the M1's revolutionary

1,500hp powerplant. It was in the area of protection, though, that the Merkava arguably started to move ahead of its rival.

As we saw with the T-64, the Cold War designers had afforded their tanks a degree of frontal protection that could never be matched in all aspects. Thick turret 'cheeks' or steeply raked glacis plates were features of the tank-to-tank gunslinger face-offs that they wanted to survive. As for the vehicle's rear or top, it just couldn't be protected with the same thickness of armour without creating a 100-tonne, 3.5m-high tank.

With the Mark 3, the Israelis adopted advanced composite armour for the first time, using it on the top, rear and flanks as well as on the front. Given Tal's philosophy that penetrations might be unstoppable but their effects had to be limited, it also ensured that all the main gun ammo, including ready rounds on the turret floor, were given improved protection.

Guy Tzur, then a lieutenant colonel commanding a battalion of the 7th Armoured Brigade as it converted from Centurions to the new Mark 3 Merkava, told me 'it was a huge change.' The Cent, even with its extensive suite of modifications, was a thing of the past; the home-made replacement 'gave me much more confidence in its protection, it was much more comfortable inside the Merkava, and it came with a very new system, the fire control system [which was] much more advanced.'

In its early wars, the IDF had shown itself able to innovate very rapidly, adjusting to the appearance of new threats or the revelation of its own weaknesses more quickly than the Pentagon's Byzantine procurement systems or those of other big industrialized nations. And no sooner had the Mark 3s been delivered than thinking started about what was wrong with them, and how they might be improved. 'There was a constant conversation between me as a battalion commander, and my team,' recalls Major General Tzur, 'with the department that built and managed the whole design of the Merkava 3. This was very good news because we could fit the tank to our needs.'

When he was subsequently posted to the defence ministry, in

charge of armoured doctrine, Tzur sat at the feet of the master, Tal, at the start of the Israeli working week. 'Every Sunday for two years I met him, and I was the client of the Merkava that he built. We were very important for him, he heard us. He gave us a lot of wisdom and professionalism.'

This constant interaction produced yet another iteration of the design, the Mark 4 Merkava, which entered service in 2005. Finally, the Israelis moved away from American-designed power-packs to the 1,500hp V12 diesel developed by the German firm MTU for the Leopard 2. The Mark 4 also took a big step forward in terms of keeping its ready ammunition safe, adopting a system pioneered by Chrysler in the US, of putting shells in an armoured box at the turret rear. This would protect the crew from any blast if this was penetrated.

The key thing that differentiated the new variant from any other tank in production at the time was the recognition that it was time to move away from a design optimized for head-to-head combat against other tanks (an evolution begun with Merkava Mark 3) to one given effective protection from threats in all axes. Inevitably this meant another increase in weight but also a reordering of existing protection.

This can be seen most clearly with the protection against top attack, for example on the turret roof. The commander's turret hatch, when open, can be seen to be around 35cm of composite armour. As with the Merkava Mark 3, great add-on packs of armour considerably increased the turret size – a measure of how the small frontal profile originally chosen to make it harder to hit in tank-to-tank fighting gave way to a much broader one. Inside these modular armoured 'wings' wrapping around the turret were multiple layers of protection separated by air gaps. Among these were 'non-explosive reactive armour', plates made of a spongy compound designed to dissipate some of the energy from the incoming metal bolt of a HEAT charge.

Although the US military had plenty of experience in Iraq from 2003 onwards in the changing role of armour and threats to it, there was no budget or political will to design what would have been a replacement for the M1 that addressed the 360-degree nature of threats through a fundamental redesign. Thus the Merkava Mark 4's protection from top attack on its turret is more than ten times as thick as the American tank's. This was to have unfortunate consequences for the Ukrainian crews sent forward in US-supplied Abrams.

In the 2000s a tit-for-tat battle between designers of defensive and offensive means remained in full swing. While the old Cold War contest between Soviet and Western scientists was over, this new one involved a struggle between countries like Israel or the US against insurgent forces using low-cost solutions while adapting very rapidly.

Instead of teams of scientists in government-funded research institutes from Chertsey to Chelyabinsk, this was the work of a few electronics nerds experimenting with components originally designed for quite different uses. A car's reversing sensor could be used to make a proximity fuse, or a hobbyist drone made into one capable of dropping grenades. And all at a fraction of the cost and time that big defence contractors took to come up with solutions.

In the skirmishes along the Lebanese frontier (Israel had withdrawn from the south of the country in 2000) and the border with the Gaza Strip, non-state militias like Hamas and Hezbollah experimented with new inventions and tactics. Inevitably, they made common cause against the 'Zionist entity', something facilitated by Iran. The Quds Force (paramilitary arm of the Iranian Revolutionary Guard Corps) assisted them with the supply of modern anti-tank missiles, know-how about enemy weak points, and ideas for making large IEDs (improvised explosive devices), as well as local manufacture of shaped charges capable of penetrating hundreds of millimetres of armour. Once again in this story, the big challenge came from the designers of these high-explosive anti-armour devices.

When Hezbollah raided Israel in July 2006, abducting two soldiers, it touched off a wider conflict, leading to heavy bombing of

Lebanon and a month-long ground incursion. The forces used in this IDF assault were much smaller than those in 1982, with correspondingly limited aims. However, the IDF made heavy weather of it, coming repeatedly under ambush, facing thousands of anti-tank projectiles.

Hezbollah's skills extended to information warfare, so it used the destruction of Israeli armour as a big theme in its messaging. One of the movement's leaders, Sheikh Nabil Kaouk, declared in a TV interview: 'Israel developed the Merkava and turned it into a symbol of Israeli military technology. It showed it off to all the armies of the world. It became a legend. And so, Hezbollah acquired weapons with the aim of destroying the Merkava. They developed. And we developed too, and we surprised them.'

The movement's TV and online news service, Al-Manar, boasted: 'Hezbollah literally massacred the enemy's Merkava tank . . . Hezbollah ambushed the Israeli armoured tanks advancing into the Lebanese villages in Khiyam and Hujeir valley, destroying 30 Merkava tanks.' As Sheikh Kaouk confirmed, in the run-up to the conflict Hezbollah had obtained advanced Russian anti-tank missiles such as the *Kornet*, *Metis* and RPG-29, embodying advanced features such as tandem warheads (used to defeat explosive reactive armour) and 'top attack' flight profiles. How real was their success then?

Israel had used 380 tanks in the brief war, all of them Merkavas, the IDF having retired the last of its foreign-made machines in 2005. Hezbollah, according to the post-operation IDF inquiry, had fired 1,200 anti-tank missiles and two or three times this number of RPGs at this armoured force. There were 45 hits on Israeli tanks, in 22 cases the vehicles were penetrated, and in five instances the Merkava completely destroyed. A further two of them were wrecked by large IEDs buried under roads. Across these incidents involving the vehicle, 30 Israeli tank crew were killed in 13 Merkavas.

To the IDF, the loss of just one in nine tanks hit by missiles compared very favourably with figures of 20 per cent during the

1982 Lebanon operation, and 59 per cent during the Yom Kippur War. It underlined the dividends reaped by the decades-long emphasis on protection in the Merkava story.

But of course, to supporters of armed factions in Lebanon or the Palestinian occupied territories, images of a blazing Israeli tank were evidence of the humiliation of their hated enemy. And this was exploited more widely across the Arab world. The international TV network Al Jazeera, in a documentary on the 2006 war, noted that the Merkava was 'a powerful symbol of Israel's technological might' that ended up 'not living up to its name or reputation'.

The experience of 2006, limited as it was, evidently caused a great deal of soul-searching in Israel about the continued value of armour. Tankers emphasized poor preparation for the operation, and ill-conceived tactical practice during it. They also moved forward swiftly with a new invention that was to be one of the most important steps in the history of armoured warfare.

Soon after the 2006 summer campaign in Lebanon, the Israeli military launched a competitive run-off between two defence contractors. For some time previously they had been developing 'Active Protection Systems', or APS, called Iron Fist and Trophy. These systems combined active radar sensors with high-speed computation or fire control, and explosive formed projectiles fired from launchers on top of the tank towards incoming projectiles. In short, the contractors' claim was that their APS could shoot down incoming missiles before they struck.

In the decades leading up to this development, tanks had been fitted with all manner of passive sensors. My own Chieftain in the late 1970s, for example, had one to tell you if someone was shining infrared light in your direction. At the time, Soviet tanks relied for their night fighting on searchlights projecting on a frequency invisible to the naked eye.

Later on, vehicles were fitted with laser detectors designed to alert the crew when somebody 'pinged' their tank; that could allow rapid evasive action, since it could be evidence that an enemy vehicle was measuring your range prior to firing. As anti-tank missiles became more prolific and capable, some armies started to look at active defences. These could transmit energy to jam the infrared or laser guidance used by some missiles.

Soviet designers had been the first to conduct field trials of these systems. One called *Shtora* or Curtain had been developed in the late 1970s, using both jammers and smoke dischargers to save the vehicle. At the same time, *Drozd*, a full-on Active Protection System in the sense that it was designed to shoot down incoming missiles, was also trialled but did not go into production.

Those who conducted these tests had to ask themselves all kinds of questions about cost and sophistication. If an APS cost $500,000, say, and was 75 per cent effective, was it worth fitting it to a $1 million tank? The fin-stabilized armour-piercing shot fired by other tanks would still get through these systems: they were an answer only to HEAT warheads. The small projectiles fired by these protection systems lacked the energy to deflect or disrupt the solid penetrator fired by a tank gun, but missiles were more fragile.

By 2010, say, nations could easily spend $5 million on a new tank. The balance of advantage was changing: the threat was very largely one from those using shaped charges; the medical costs of treating seriously wounded crew were rising steeply; and computational advances increased the chances of successful interception at speeds which meant no human was in the loop.

It was indeed in 2010 that, having won the IDF's competitive process, the Trophy system entered service. Its price was quoted as 30 per cent of the tank's value. The following year, following the firing of a Kornet from Gaza, a Merkava Mark 4 successfully blew the incoming missile out of the sky.

The decision was taken not just to roll out the successful protection system on tanks but to use it on personnel carriers like the Namer too. The Israelis thus became the first army to deploy at scale an effective system for intercepting all manner of threats. Missiles, recoilless rifles and RPGs could all be neutralized. And when one added it to the Merkava's composite armour array, they had changed the calculus in armoured warfare.

As this happened, quite paradoxically, Israel all but lost faith in the tank. The 'father of the Merkava', Israel Tal, died in 2010, having ceased for some years to be influential in the debate due to illness. He had argued that tanks were 'the decisive strategic weapon', vital to winning his country's wars. In the wake of the 1973 war, the fleet had been built up to almost 5,000, as we have seen. But as the older models had gone out of service, the cost of replacing them escalated and in 2013 the decision was made to halt tank production. By 2020 the standing armoured corps was operating just 480 machines, with around 700 in store for reserve brigades, all Merkavas. Doctrine had evolved, and they were no longer central to the IDF's way of war.

As a major general in charge of ground forces, Guy Tzur explained the shift in thinking after the 2006 war thus: 'when you are a military force and you have a disappearing enemy, then your challenge is that your enemy will see you before you see him, shoot before you can react, then disappear before you eliminate him. This is the big challenge that we understood, and how hard it was, after 2006.'

Major General Emanuel Sakal, who followed a similar career path, suggested that the change was about embracing combined arms in all aspects: 'you can't win alone, you can't fight alone . . . the secret of ground warfare is working together.' It might be argued that this was obvious even during the Second World War, but the IDF had been on a journey in which, following 1967's 'perfect blitzkrieg', the armoured corps had enormous influence. Along with the focus on non-state enemies and the low chance of any tank-to-tank fighting, or indeed large-scale exploitation involving advances across hundreds of miles, by the 2000s it was time for this organization to change.

The smallest building block for tank units is the platoon, most commonly four vehicles. At times in the Second World War some armies went for five or even six, since they wanted the unit to remain effective even if it lost one or two. Some others, like the Soviets, used three tanks, since their focus tended to be on the next level up the chain of command, a company of ten machines (with three platoons of three each and a company commander's tank).

As the IDF's regular tank fleet dwindled below 500, it adopted a two-tank platoon. Their logic was that the vehicles should never be committed singly, since one must always be able to cover the other when it moves; that was therefore the lowest level of support for an all-arms unit. Larger groups, companies, could be used if necessary, but what this change really announced was the relegation of the tank to a support weapon.

It was still needed, because Israel had developed such big, heavy armoured personnel carriers, but these vehicles could only be lightly armed. They were so enormous, even without a large weapon-bearing turret on them, that the Merkava was vital to escort them. The key point of difference between a Merkava and the Namer APC is its gun. It is powerful enough to pierce another tank's armour by sheer force of kinetic energy, with a range of alternative shells for lesser enemies. It's the ability of the ground force 'to have firepower of its own', to quote Major General Sakal, rather than depending on the 'arrogant', distant air force, that is one of the main things keeping the tank alive.

With its powerful armament, thick armour and effective sensors, the Merkava's mission became one of backing up soldiers even in the most difficult places, including the nightmare of an earlier generation, street fighting. Visiting exercises at the Tze'elim training ranges in 2013, one military correspondent talked to units engaged in training to assault built-up areas. 'The armour has adapted itself to the developing battlefield and to guerrilla warfare,' said Lieutenant Colonel German Giltman, a deputy armoured brigade commander, describing the prospect of an engagement at that time against another tank as 'a dream'.

Another officer at those exercises, Lieutenant Colonel Isham Ibrahim, told the writer: 'Today, as an armoured battalion commander, I'm looking to get into the city. I want to be in the urban area.' This wish would be fulfilled during numerous skirmishes with Palestinian armed groups but ultimately in the major actions following the Hamas assault on southern Israel of the 7th of October 2023.

In time, comprehensive histories of these horrendous events will appear. The question of what Hamas hoped to achieve with its campaign of murder, rape and kidnapping may then be put in some sort of context. Similarly, Israel's motivation in sending armoured divisions into one of the most densely populated parts of the world, causing such a tragic loss of Palestinian civilian life, will take time to become clearer, as will questions of whether its commanders will ever have to justify their actions in court.

With our narrow focus, it suffices to say that the military aspects of Hamas's 7th of October attack, i.e. the defeat of the IDF on Gaza's borders, was a well-planned and, in its own terms, highly successful operation. Twenty-three Israeli military positions were overrun in carefully coordinated complex attacks, and the IDF had more than 300 soldiers killed as well as dozens captured.

The psychological blow from this surprise attack, falling on the anniversary of the 1973 war, was of comparable intensity to that earlier one. From a militant Palestinian perspective, images of people flag-waving atop a disabled Merkava at the separation fence became one of the most powerful emblems of their success.

During the actions of the 7th of October, a couple of IDF tanks were hit by drone-dropped munitions and, combined with the setting fire to some captured later, ten Merkavas were put out of action, as well as 24 Namers and 22 other APCs. Only two, though, of those are thought to have been completely destroyed. When Israeli armoured brigades rolled in a couple of weeks later, beginning months of operations in Gaza, many predicted they would be sucked into an urban graveyard. Hamas had widely distributed a locally made RPG round designed (with Hezbollah assistance) to defeat both explosive reactive armour and the Merkava's composites.

The *al-Yassin* 105mm rocket has a tandem warhead – it has two charges. It is designed to give its target a one-two punch. The first triggers reactive armour, and the second, a fraction of a second later, is then intended to penetrate. Hamas also fielded a smaller number of *Tharallah* systems, a Hezbollah innovation using a pair of Kornet missiles, the first to trigger a Trophy system, the second to hit the tank before it can acquire a new target.

This weapon may have been used to hit a Namer APC a couple of days into the operation, killing 11 Israeli soldiers. But this was an isolated success for the group. While Israel has kept the losses of armoured vehicles classified, and Hamas has certainly succeeded in knocking out a few, there are signs that the Israeli tactical approach and armour protection systems worked very well to prevent many losses. Videos of Hamas attacks, mostly with al-Yassin rockets, have been released to incite the faithful but the great majority freeze-frame at the point of an explosion. Indeed, meticulously researched open-source investigations of 194 strikes of Merkava tanks between November 2003 and August 2024 came to the conclusion that just two of them had been completely destroyed, both losses attributed to massive IEDs made from unexploded Israeli air force bombs that had been salvaged by Palestinian fighters. The same analysis suggested that just two of 63 filmed strikes on Namer heavy APCs had destroyed their targets. Adding to this impression, Sascha Bruchmann, of the International Institute for Strategic Studies, wrote: 'An analysis of the data of killed and wounded Israeli soldiers shows only a few that may have been tank crew.'

During a previous, more limited, round of fighting in 2014, the IDF had estimated it could lose 300 soldiers fighting its way into Gaza City; in the event, during November 2023, it was a couple of dozen. There were many ingredients in this, not least using a level of force that Western armed forces would have regarded as unacceptable, but Israeli generals drew the conclusion that both their armoured developments and doctrine had been vindicated.

What is most noteworthy about the Israeli approach over the past two decades is the way it has diverged from that of the US and other Western countries. None of them deploy heavy personnel carriers in the 50–65-tonne range. At the time of Russia's 2022 invasion of Ukraine, none of them deployed active protection systems on their tanks, though they have since scrambled to rectify this, with many buying the Trophy system.

Looking at more fundamental elements of their tank design, they have been far slower than the Israelis to react to the 360-degree nature of current threats, and with it the matter of how to up-armour the vehicle's weak points. In some ways that's because they faced lesser challenges and in others because their military-industrial complex was less nimble.

And if success is measured in large numbers and plentiful exports, the Americans outdid the Merkava with the M1 Abrams, proclaiming it with equal fervour to be the 'best tank in the world'. And at the time of the American tank's first combat, it's easy to see how that claim was justified.

M1 Abrams

Entered service: 1980
Number produced: nearly 10,000
Weight (M1A1): 67 tonnes
Crew: 4
Main armament: 105mm gun, 120mm on later models
Cost: $2.8m in 1982, $24m in 2022

M1 Abrams

It was mid-afternoon on the 26th of February 1991, about 75 miles south-west of Basra. The US Army's 2nd Armored Cavalry Regiment was advancing across the Iraqi desert to make contact with troops from Saddam Hussein's Republican Guard. But the morning mist had given way to a sandstorm, visibility was just a few dozen metres, and they were moving at little more than a walking pace. It was a deadly game of blind man's bluff in which technology and training would decide the issue.

Although the desert here is very flat, there are gentle rises and dips, often barely visible to the eye, that offer some tactical possibilities. Peering into the sandy murk, the cavalry crews in their Bradley Fighting Vehicles had an advantage – thermal imaging sights allowing them to see further than the tanks, and to see hotspots: Iraqi soldiers running from their trenches to man their vehicles.

The commander of 2nd ACR's Eagle Troop had already engaged some enemy soldiers among houses. At around 4.07 p.m., with enemy armour appearing to his front, he ordered his M1 Abrams tanks to move ahead of the Bradleys, ready for action. The troop had two platoons, each of four, plus his own Abrams, making nine in total.

Trundling forward, these machines represented the state of the art in tank design for the late Cold War. Weighing around 67 tonnes when fully loaded for combat, with anything up to 600mm of frontal armour and a 120mm gun, a gas turbine engine produced the power to propel such a huge vehicle nimbly across the sands.

Eagle Troop, about 150 soldiers in total, continued its advance. It drove onto ground just a few metres higher than the surrounding desert, Captain H. R. McMaster in command. A force of nature, he would later be one of the few officers to emerge from the post-2003 occupation of Iraq with distinction, going on to serve at cabinet rank as the president's National Security Advisor. But at this moment McMaster was just a captain, albeit one of fearsome focus and organizing abilities. Leading Eagle Troop on, he watched as an enemy defensive position was unmasked: 'As our tank came over the crest of the imperceptible rise . . . Sergeant Craig Koch, the gunner, reported "tanks direct front," I counted eight T-72s in prepared positions. They were at close range and visible to the naked eye . . . All nine [US] tanks began engaging together as we advanced. In approximately one minute, everything in the range of our guns was in flames.'

Iraq's invasion of Kuwait had produced a wider war in which America's M1 tank received its baptism of fire. On the 26th of February, questions that tank soldiers had been asking for years, about the relative merit of the latest Western machines versus the T-72, would be answered emphatically. A few minutes after McMaster unleashed his tanks, about 7 miles to the north, another of his regiment's sub-units, Ghost Troop, was advancing on a similar west-to-east track.

Ghost Troop's commanding officer, Captain Joe Sartiano, whose scout Bradleys had already engaged in sporadic contacts with outlying Iraqi units, heard the uproar to the south. Here too, gradually gaining a little height on a near-invisible ridge opened up an enemy position – about a dozen Iraqi machines consisting of BMPs or infantry fighting vehicles and a few tanks. Like much of the Iraqi army that had weathered weeks of airstrikes, they had been driven into pits dug by bulldozers to protect them from the bombing.

Sartiano's gunner used his laser sight to get an instant reading on the BMP to their front; it was 548 metres away. He fired and hit it with a discarding sabot armour-piercing shot, then loaded a HEAT shell to take on a second BMP. Like McMaster, Sartiano ordered tanks front, and his 2nd Platoon, M1s, rolled forward. The burning BMPs were adding acrid black fumes to the sandstorm, and Sartiano watched uneasily as 1st Lieutenant Andy Kilgore with his four M1s vanished into the murk.

Kilgore recalls: 'when I poked through the smoke I realized that we had a company sized defence spread out in front of us. I told my gunner to engage.' A gunner in one of his tanks pressed the button to lase the range to an Iraqi vehicle in front of him: it returned zeroes, indicating less than 200m. The enemy, he says, 'was never more than 2–300 metres away'.

The M1s' 120mm guns barked away, sending shells flying, brewing up BMPs and tanks alike at close range. Kilgore's platoon had gained a complete surprise. There was no return fire from the vehicles – 'Our battle was a slaughter,' he explains; 'we were shooting vehicles that weren't cranked, they weren't running, we co-axed the crews running towards them.' ('Co-axes' are the machine guns mounted next to the main gun.)

Sartiano, some way back, could hear the thunderclaps of 120mm gunfire but was unable to see what was going on in the smoke. He bellowed to Kilgore over the net for a report: 'Answer the god-damn radio, over!' But the lieutenant had his hands full. They had made swift work of the Iraqi vehicles, now the surviving Republican Guard infantry rushed forwards, firing with small arms and anti-tank weapons. He and the other commanders blasted away with the heavy machine guns fitted to their cupolas, chopping down the assaulters: 'body parts were flying everywhere.' Their gunners used co-axes to pour on even more mayhem.

Iraqi survivors kept rushing onwards. One Abrams commander, standing in his hatch, drew his pistol and fired down from his lofty position. Kilgore realized they might soon get boarded by desperate men with grenades, so he got on the radio: 'Back up! Everybody back the fuck up now!' The four tanks kicked into reverse and sped back into the smoke, reappearing to Captain Sartiano's front. Their push forward, claiming a dozen armoured vehicles and scores of their infantry, had lasted just three minutes.

Sartiano assembled his troop on the low feature, 'scrambling' the Bradleys and M1s into mixed groups with two vehicles of each kind. In this way, the Bradleys were able to use their superior thermal imaging to look out for targets.

To the south, meanwhile, McMaster pressed through the first position he had assaulted, and a few miles to the north-east discovered another company of T-72s (about one dozen) deployed in a circular position, probably some kind of reserve. Advancing, the Eagle Troop armour went through a minefield without loss; as the anti-personnel devices went off, 'they sounded like microwave popcorn popping', McMaster would recall later.

Quickly, they closed the few hundred metres to the Republican Guard position. They were so close, McMaster recalled, 'I looked at the enemy tank commander, he looked over his shoulder at me, I could see the expression on his face, and we engaged that tank at very close range. You couldn't tell the difference between the enemy tank being hit and our gun going off, and big hunks of metal and sparks arced right back over our heads.' Eagle Troop's battle had lasted 23 minutes, in which 40 tanks and 13 BMPs were claimed destroyed without loss on their side. Established on this second position, they took hundreds of Iraqi prisoners.

To the north, Ghost Troop was arrayed on top of the rise by around 4.40 p.m. There followed a series of engagements as Iraqi company groups (of 10–20 armoured vehicles each) tried to move across the front of Ghost Troop's position. 'People were screaming and hollering on the net, saying they were coming in waves, but they were escaping,' Kilgore told me. Without realizing, they had got across one of the Republican Guard's escape routes from Kuwait, he believes. 'It was not a formed counter-attack, but we were in their way,' he says. Seeing a column of Iraqi vehicles moving around 3,500m to their north-east, possibly trying to break contact, their artillery liaison officer called in a barrage of 155mm shells and artillery rockets onto the target.

Sky-lined as they were, Sartiano's troop came under effective Iraqi fire for the first time. The Iraqis had called down an artillery barrage on them. Then there was another problem. Some surviving Iraqi crewmen jumped into one of the BMPs previously hit by Kilgore's platoon, opening fire with its 73mm gun. A Bradley, Ghost 16, was hit by a shell, the gunner, Sergeant Nels Moller, shouting 'What was that?' Moments later, a second round hit, striking the box next to the turret that held its anti-tank missiles, causing a large explosion that killed Moller and wounded two other members of his crew. Seeing what had happened, Sartiano, who was a little to the rear, sent a 120mm sabot round towards the BMP, knocking it out. Lieutenant Kilgore, who'd been out of his tank fetching water from the turret rear stowage, was knocked off the vehicle by the shockwave as the metal dart fired by Sartiano's gunner flew past.

As the late afternoon gloom deepened, Sartiano started to worry that his Bradleys had fired off most of their TOW missiles – the most effective long-range tank-killers he had. But around 7 p.m. another Iraqi company was detected moving towards them. One of the Bradley commanders, shaken by the suicidal Iraqi advances, told a reporter from *Stars and Stripes* soon after it happened: 'We just couldn't understand it. I still don't understand it. Those guys were insane. They wouldn't stop.'

An M1A1 assisting those Bradleys picked out an enemy vehicle at 3,700m with its thermal sight. Letting fly with a HEAT round, the American tank knocked it out. Even if they had been able to see the Americans, the Iraqi T-72s were quite unable to match the Bradleys (firing TOW missiles) or Abrams at ranges over 2,500m.

The Battle of 73 Easting, as 2nd ACR's action became known, ended that evening as other units pushed on to complete the Republican Guard's rout. An early After Action Review suggested the troops engaged on the 26th of February had in just a few hours destroyed dozens of vehicles, killing hundreds of troops as well as capturing more than 500. But the wider engagement in this area, over two days, claimed something like 160 Iraqi tanks and a similar number of other armoured fighting vehicles. The 3rd Republican Guard Mechanized Division (also known as the Tawakalna Division) managed to save only two dozen or so of its 200 tanks during the battle.

And the cost? One Bradley was destroyed (Ghost 16 being finished off accidentally by an American tank that wrongly identified the smoking wreck as Iraqi) and an M1 tank also put out of action by 'friendly fire', a cluster munition from one of the 155mm shells fired towards the Iraqis.

The superiority of the M1 had been dramatic – even if its performance had not been flawless. Two of Ghost Troop's tanks, for example, had suffered failures with their gun and electrical systems. In the grand scheme of things, though, such teething problems did not diminish the dramatic disparity in tank losses across the whole operation. The Americans had just over 2,000 M1s engaged, out of a total of 3,000 tanks across the Coalition.

The number of Iraqi tanks *destroyed* was a similar number in the thousands, though a great many of these had been taken out during the weeks of air attacks before Coalition forces crossed into Kuwait and Iraq. There was also a great deal of over-counting, Andy Kilgore pointing out that many Iraqi targets at the 73 Easting were engaged over and over: 'multiple people will see a target, engage it, and claim the kill.'

By contrast, the M1 tank loss tally for US forces during their 100-hour ground war against the Iraqis was 23, of which nine were completely destroyed. There were numerous friendly fire incidents, with just three or four of the M1s being knocked out by Iraqi tank fire. It was a stunning display of combined arms warfare, the continued value of large tank formations, superior training and of course technological superiority.

Why hadn't the Republican Guard T-72s made a greater dent in the Coalition's armoured phalanx? During the Battle of 73 Easting the M1 and Bradley's thermal sights put the Iraqi crews at a grave disadvantage. The Iraqi commanders could not see into the sandstorm or smoke, so the emergence of American armour came as a terrible surprise. Added to this, Kilgore's platoon sergeant, Waylan Lundquist, says there was a big difference at the human level: 'training-wise they weren't at our level.'

There were other issues too. The tactical situation, with the

Guard trying to withdraw across the front of 2nd ACR, broadside-on, also made it an unequal fight. As for the stationary Iraqi vehicles initially encountered, many could not engage nearby targets from the deep pits bulldozed into the sand to save them from air attack. Quite a few were hit by the Americans as they reversed out, unable at that moment to return fire.

Analysts were also at pains to point out that the T-72s used by Iraq (many of which were assembled in the country) were an export version. Their armour, optics and shells were not of the same quality as those used by the Soviet army. Even so, as the US General Accounting Office reported shortly after the war, 'several M1A1 crews reported receiving direct frontal hits from Iraqi T-72s with minimal damage.' The Abrams' thick armour had been sufficient to protect them from what was still a formidable 125mm gun.

The M1 was not the first tank to use a gas turbine engine, nor to have thick composite armour, but it did manage to exploit America's prowess in electronics, microprocessing and optics to an impressive degree. It fused the cold steel that symbolized the industrial age with new inventions emblematic of the information era.

Today the processing power of an M1 fire control system, or the technology of its radios, might be looked upon as antediluvian, charmingly basic. Even at the time, though, the M1 and its Nato cousins the British Challenger and German Leopard 2 were regarded as the first tanks in which the cost of electronic systems exceeded that of the basic machine with its armour, gun and engine.

This was in defence terms inflationary, since the integration of these various electronic systems threw up untold bugs and glitches that required thousands of hours' labour to resolve. It meant that armies seeking to replace tanks from the 1950s and 60s could afford fewer, and had to wait longer for them. To give just one example: at the time the M1 went into production, each thermal imaging sight was reckoned to cost hundreds of thousands of dollars, about one quarter of the entire price of the vehicle. Yet the desert battles of 1991 were to show that it could confer a decisive advantage.

Finding targets through these advanced optics, communicating securely, and the use of advanced fire control systems giving tankers at last a genuine ability to fire just as accurately on the move as they did when static were all ingredients in the M1's success. It also had something that denizens of the Sherman in its desert days would have remembered fondly: terrific reliability. When the 2nd ACR redeployed several weeks later from southern Iraq to a base in Saudi Arabia, a distance of 600 miles by road, its 120 or so Abrams suffered just a couple of breakdowns.

The M1's origin story represents in many ways a typical saga of Cold War arms development. It had a long gestation fuelled with frustrating Allied wrangling, cost increases and political criticism from Capitol Hill. That it turned out as well as it did was due in no small part to corporate America's management skills, so vital to the Sherman's success, that still lived on in the 1970s at the Detroit Tank Arsenal and the Lima plant in Ohio.

Where does the M1 story begin? You could argue that it was as far back as 1961 when early concept studies began on a vehicle that would be quite different to the Patton series, developed by evolution from the late 1940s to the M60 in the early 60s. The Americans soon started talking to the Germans and by 1963 a joint project called Main Battle Tank 70 or MBT70 was underway.

The ambitions for this new vehicle were not unlike those that Morozov had for the T-64, development of which was proceeding just ahead of the new Nato tank. The MBT70 was also designed around a three-man crew, replacing the human loader with machinery, reducing internal volume and the vehicle's height at the same time. So determined were they to squash its silhouette down that the experimental tank removed the driver from the hull and sat him in the turret. Just as tanks had gyro-stabilized guns by this time, so the driver, sitting in a motorized capsule, would be kept pointing in the direction of travel even if the turret was traversed off to the side or rear.

As trials using several prototypes proceeded during the mid-1960s, reports came back of seasick, disoriented drivers. Differences between the partners quickly multiplied. To complete the farce, American and German teams argued over whether to use metric or imperial sized components, and over which main armament to fit. The Americans wanted a hybrid 152mm gun/missile launcher, the Germans a 120mm gun. The latter weapon survived the project debacle, being developed into the main armament for the M1 as well as the German Leopard. Having gone their own way, to the chagrin of a key Allied government, the Pentagon advocates for an all-American procurement would find the Leopard coming back into the picture later in the story.

With politicians in Congress focusing on a unit price of $1m a tank, funds were shelved and in January 1970 the US pulled out of the project. The Germans went in a quite different direction while an American cut-price version of the revolutionary vehicle stuttered on for another year as the XM803.

In many ways the MBT70 was a typical Cold War weapons saga: as armies shrank, the Allies looked for partners to share ballooning development costs; but trying to reconcile different nations' needs caused still more inflation; and an initiative intended to serve as a model of Allied cooperation fell into recriminations, bitterness and the cancellation of the project with $400m spent and nearly a decade wasted.

Given the pace of competition with the Soviet Union and time already lost, the Pentagon did not want to dawdle and by 1972 was examining new concepts for a future tank. These initial studies looked at different sizes of vehicle ranging from 49 to 57 tons, the bigger version having an estimated price tag of $500,000 per tank.

The following year, General Motors and Chrysler were given contracts to develop prototypes, the latter effort being headed by an engineer called Philip W. Lett. He was an Alabaman who'd been working in armoured vehicle design since 1950, and who would earn the title 'father of the M1'. Lett had worked at the Detroit Tank Arsenal during this time as an engineer on the Patton tank projects. Like his predecessors at that plant, he put a high emphasis on reliability and adaptability. Lett understood from the outset that pitting these two big corporate names against each other would result in 'a fierce competition for big stakes'. Indeed, the Chrysler management well understood that if they lost out to GM it would likely spell the end of their armoured vehicle division, and vice versa.

The question of cost increases – and these were years of high inflation in the civilian economy as well as defence – meant that focus returned time and again to the price of this new project. What Western countries discovered, as their weapons grew more sophisticated and therefore took longer to develop, creating higher costs spread over smaller production numbers, was that inflation in defence was significantly higher than that in the wider economy.

In 1942, a Sherman had been specified to cost $35,000. Adjusting that for the growth in prices, it would be $90,000 in 1972 (or $675,000 in 2024). Tank inflation had of course been significantly higher. As with Britain and the Centurion procurement, the US had paid several times as much in the 1950s as it had a decade before. By 1961, an M48 Patton, for example, cost the Pentagon $309,000. Developing by evolution, costs had been contained on its successor, the M60. As the decision about taking a bigger leap forward in the early 1970s crystallized, a late-series M60 could still be bought for less than half a million dollars.

Les Aspin, a Democratic Party congressman from Wisconsin, was a frequent critic of military procurement. He had urged the cancellation of MBT70 when its price edged towards $1m a copy, saying 'no tank is worth that money.' By 1973, with estimates of the successor 'cheaper' design creeping over $900,000, he compared its cost to the M60 tanks recently bought, arguing: 'I am sure that the Army's new tank is not twice as good as what we have today.'

While the two big industrial contenders for what was at the

time called the XM1 contract stepped back from many of the MBT70's innovations (for example, the autoloader was abandoned, taking the crew up to four again), Lett's Chrysler prototype would involve some important innovations. The most important of these was its gas turbine engine. Lett was an enthusiast for this form of propulsion. Chrysler chose an Avco-Lycoming engine adapted from use in a helicopter and capable of producing 1,500hp to power its prototype.

Given that diesel engines could produce similar figures, what was the attraction? Lett argued: 'its inherent reliability – it had 30% fewer parts than the diesel, and it offered 2,000lb net weight savings that allowed additional armour protection to be placed on the vehicle.' The designer believed that reliability would reduce maintenance. The later experience of crews with the Sisyphean task of cleaning its air filters would have provoked hollow laughs and expletives to that latter point.

There were other arguments in favour of the gas turbine: that it was quieter, smaller and lighter than the equivalent, giving a reduction in the volume of the engine bay. And it did have an advantage in horsepower, because, whereas around 160hp of a 1,500hp diesel was sacrificed to cooling it, the figure was just 30hp with Lett's chosen turbine.

Inevitably, though, there were trade-offs, most importantly the gas turbine's fuel consumption. In typical test conditions the Abrams sucked down 1.67 gallons of fuel *per mile*. However, in some circumstances (for example, cross-country in heavy mud) it could go far higher, burning almost twice as much gas in competitive trials run by the Swedish army. Comparisons might have been made to the Centurion's very thirsty Meteor petrol engine – but of course the Avco-Lycoming plant produced more than double the horsepower.

The answer was to give the M1 a large enough fuel tank (300 gallons) to allow it to travel 200 miles at a steady road speed. But even so, the engine's thirstiness proved to be a logistical headache, and many years after introducing the M1, the Americans would reach the same conclusion that the Centurion's designers had at the outset: that fitting an additional motor, an auxiliary power unit just to run its electrics, could allow the main engine to be switched off rather than idling while the tank was halted, giving big savings in fuel.

The M1s used in the Battle of 73 Easting didn't have auxiliary engines, so to save fuel whenever possible the crews switched off the gas turbine, running their electronic equipment off batteries. But putting an engine originally designed for a helicopter into a tank caused other challenges. Andy Kilgore, who later retired as a lieutenant colonel, recalls: 'Our biggest issue with maintenance was the massive amount of air that engine needs and we were constantly cleaning air filters.' His former platoon sergeant Waylan Lundquist chimes in: 'It's an oxygen pig, it just needs way too much air, and the other problem is that it sucks so much fuel . . . it's tied to a fuel truck.'

Once in service, the Abrams became the tank equivalent of the iPhone. It could do all sorts of amazing things, but the user constantly worried whether it had enough juice.

Back in the mid-1970s, as the project went on, there were inevitably moments when new challenges fed the doubts of those wishing to kill the project. The initial destruction by Egyptian anti-tank missile teams of Israeli Pattons during the Yom Kippur War triggered that recurring worry about whether the tank was still viable on the battlefield.

So, as the engineers worked away on the practical challenges of eking out extra performance or limiting weight, the Pentagon hierarchy went repeatedly into battle on the larger plane, defending their case for why they needed this new war machine at all. The General Accounting Office, often scathing about military procurements, flagged up the spread of anti-tank missiles. And a powerful voice among those holding the purse strings, the Senate Armed Services Committee, noted in 1976 the 'need to re-evaluate the role of the tank on the modern day battlefield'.

Charting the tank journey from the Mark IV or Renault of the First World War, we can see how vehicles that took just a few months to evolve from wooden mock-ups into production prototypes had become systems like the M1, which, even excluding the MBT70 period, could take eight or nine years to be developed from the blueprint stage to entering service. Naturally this was a consequence of the enormous complexity of these later machines, assembling a dizzying number of mechanical, composite and electronic components, as well as so many lines of software code. And this ever longer journey from concept to delivery raised the political jeopardy for these projects.

For those backing the XM1 this familiar Capitol Hill obstacle course – with sceptics on one side saying it was yesterday's weapon, and on the other that the Soviet Union had something better – was all part of the game when billion-dollar contracts were at stake. And at this stage in the 1970s there were plenty of uncomfortable revelations about what their adversary was up to. Once Western engineers were briefed on the armour-piercing powers of new Soviet tank guns, the protection required for XM1 increased, with the inevitable consequences for its weight. This seemed to underline the wisdom of choosing the gas turbine but posed fresh dilemmas about weight and cross-country mobility.

In the search for improved mass efficiency, both in its own armour and the projectiles it would use against the new Soviet armour arrays, the Americans adopted an exotic, politically problematic substance. Engineers knew that uranium was one of the densest materials, so were persuaded by the idea of using the waste metal left once the radioactive isotopes had been extracted from it. This 'depleted uranium' (DU) was therefore not radioactive and could be employed both in armour-piercing penetrators and in elements buried within later versions of the M1's composite armour. The Germans were unhappy with this development, pointing among other things to the poisons produced when DU penetrators travelling with enormous energy struck armour.

In the UK they were phlegmatic about inflicting this on their enemies, but chose not to take the risk with their own people of incorporating DU in the secret armours that were emerging from the research at Chertsey. The Americans had learned of Britain's plan to develop new composite armour, codenamed BURLINGTON, in 1973 and felt they could go even better by using DU. Although the new protection was more efficient for its weight than steel, the XM1's mass inevitably grew, which in the long term favoured Chrysler's gas turbine-powered machine.

Chrysler's chief engineer, Philip Lett, built other survivability features into that firm's prototypes, most importantly a different philosophy on ammunition storage. Learning the lessons of the Second World War, in which ammunition fires had been such a recurrent feature in tank losses, Soviet and British designers stored rounds lower in the tank, below the turret ring. The idea was that since that upper part of the tank was exposed during fighting, it took more hits, making the danger of explosion higher if you kept the rounds there, lower if you put them in the hull.

However, Lett's approach with the new American tank was to keep enough rounds to fight an engagement in that rear part of the turret, called the bustle, that projects over the engine, but to protect them far better. 'For the first time in the history of the tanks,' he wrote, 'a new tank embodying armor within the tank was produced, which separated ammunition and fuel from the crew.' In the turret, Lett's solution was that after the loader pulled each round from its rack, an armoured door closed, separating the people from the ammo. If it was hit, and shells ignited, there were panels in the turret roof that would blow out, venting the blast upwards, rather than forwards towards the crew.

Taken together, the use of composite armour, armoured fuel tanks, secure ammunition storage and new fire suppression systems were designed to guarantee the vehicle's safety. Richard Hunnicutt, the leading historian of American armour, argued: 'crew survivability had the highest priority on the list of characteristics for the new tank.' There are parallels with the Merkava here, of course, with the armies of the late industrial period being so sensitive to casualties.

Lett would also design in the ability to upgrade from a 105mm main armament on the prototype to the 120mm the Germans had developed for the MBT70. This would prove important, not just because it gave the XM1 growth potential, like the Panzer IV or Centurion designs had, but down the line it would help to mollify some of the tensions between the US and West German governments that the cancellation of that earlier project had left.

The other quality that Chrysler wanted to build in was reliability. One way of doing this was to reduce potential points of failure. Lett explained: 'from the very outset of the design program a strict control was placed on the number of types of mechanical fasteners, hydraulic fittings and electrical connectors that could be used on the vehicle.'

During 1975 and early 1976, the Chrysler and GM machines spent months roaring around the test tracks at Aberdeen Proving Ground in Maryland, firing rounds downrange and testing various armour arrays to destruction. At a late stage in this process a prototype of the German Leopard 2 was also brought into the trials.

In November 1976 the outcome of this competitive process produced uproar in Washington, DC and Bonn, seat of the West German government. For choosing a winner required the Pentagon to make a loser out of one of those two American industrial giants, each backed by its supporters on Capitol Hill, and once that had been done, to broach the ugly truth that America was never going to give a multi-billion-dollar order to the Germans when its own industrial interests were so deeply engaged.

In the first instance, matters were made even trickier because by the summer of 1976 the army was gravitating towards the GM prototype. But there were many voices who preferred the Chrysler, not least in the upper reaches of the Pentagon. Matters came to a head in July 1976 when an army decision to go with GM was overruled by Donald Rumsfeld, the Defense Secretary. He would later write that the affair was a textbook case in how the 'iron triangle', that alliance linking military, Congress and industry, tried to dominate America's lucrative defence procurement business. Rumsfeld, not long in the job, believed 'the Army had no doubts its position would prevail. Its leaders seemed to assume my role in the decision would be to approve their recommendation.' They had even sent out the press release announcing that GM had won. In his later memoir, Rumsfeld remarked tartly: 'I was notably unhappy about being put in that impossible position.'

So he held up the decision, got the army to recall its press release and listened to some other Pentagon officials who convinced him that Lett's Chrysler design with its gas turbine 'would be more agile and efficient'. That November, 'Rummy' ruled in favour of Chrysler, giving them an order to make 11 prototypes. Inevitably, there was grumbling from quarters of the military and those congressional districts where a GM win would have guaranteed thousands of jobs.

Having taken this political pain, it hardly seemed likely that the Pentagon would then go on to choose the Leopard. Alas for political expediency, that German machine was proving embarrassingly good in the tests at Aberdeen. 'United States Army officials acknowledge,' *New York Times* correspondent John W. Finney reported in March 1977, 'the Leopard, particularly when it is firing on the run, demonstrated a 30 percent greater probability of hitting a target over the XM-1. According to a still-classified test report prepared by the Aberdeen test center, the Leopard also demonstrated greater range, less fuel consumption and better reliability.'

In the end, the Pentagon rejected the Leopard on the grounds that it would cost 20–30 per cent more than the $922,000 per tank being quoted at the time for the Chrysler vehicle, and that it was heavier. The only comfort given to the West Germans was that the American tank would switch to their 120mm gun design (but made in the US, of course) further down the line.

Chrysler's commitment to that price tag was typical of the

tactics used by US defence contractors to win an order. Inevitably, the American tank would swell up both in cost and mass as the project went on. A main contractor that had quoted low in order to get the business would use so-called 'get well' contracts, charging a premium for any improvements made by the customer, in order to balance its books.

So, when Chrysler's early development contract committed them to build 465 tanks at the $922,000 price tag, people in the company doubted whether it could be done. Chief engineer Philip Lett's worries hardly went away when his firm beat GM. Far from it. He later acknowledged that their initial pricing involved 'a risk of huge proportions', given all the development snags that might crop up. In 1982 the business was bought up by General Dynamics, Lett shifted to the new employer, and the Pentagon, once committed to the design, ended up paying far more per tank.

However, back in the late 1970s the whole project was still fragile. Those in the intelligence world, reviewing the latest assessments of Soviet anti-tank rounds and armour, wondered, in their highly classified reports, whether the XM1's vulnerability to these developments might not expose it to fresh criticism in Congress or even cancellation. In a secret memo from 1978, the UK Ministry of Defence's Chief Scientific Adviser told staff not to share their classified reporting with US colleagues, and 'emphasized the sensitivity of our new assessment in the US at this critical moment in the development and procurement programme of XM1'. Rumsfeld's successor as Defense Secretary had 'admitted the XM1 programme was vulnerable'.

But the project survived and in February 1980 the first production tanks were delivered. The vehicle was named after General Creighton Abrams, who had commanded an armoured battalion during the Battle of the Bulge. In February 1981 the main production order was given – for over 7,000 vehicles. As for the 'million-dollar tank', an idea that horrified some congressmen in the early 1970s, by 1982 the cost of those mass-produced M1s was coming in at $2.8 million each. Those who played the Washington procurement game were as adept at cooking the cost estimates to get a contract as they were test results, and by this point the Abrams had achieved unstoppable momentum.

While the USSR was building many more tanks per year (M1 production peaked at 840, whereas the Soviets built around 3,000 annually at that time), this order pushed the American tank industry to its biggest production effort since the war. The government-owned Lima plant took the lead building the Abrams, but the US Army was concerned that with its requirement for more than 7,000 of the vehicles, a planned build rate of 30 a month was insufficient. So production was started at the Detroit Tank Arsenal too, with the result that during the 1980s the US was making several hundred tanks each year. Of course this rate, peaking at 996 in 1987, was still low compared to the glory days of 1943, but it was to prove the last great tank-manufacturing campaign in any Western country – at least until the time of writing.

By 2000 they had built nearly 10,000 Abrams, a figure that included exports. Of this figure, 6,306 were built at Lima and 3,442 in Detroit, totals that recalled the Sherman days, even if they were achieved over a longer period. Well before this, the collapse of the Soviet Union in 1991, with resulting deep cuts in the US military, had changed matters radically. Not only did they need far fewer of them, but questions about the utility of tanks were being asked once more, and the Americans had adopted an innovative system of fleet management.

The smaller military only required around 1,700 Abrams. There were another 700 pre-positioned in various parts of the world, and more than 3,000 unused, kept in storage. By the 2000s, the American system became one of rebuilding obsolete M1 tanks, often taking them back to bare metal, fitting them with the full suite of new innovations, then returning them to service, rather than buying a new design of vehicle.

Thus in 2003, when that same Donald Rumsfeld who had signed off the M1's production in 1976 had returned to the Pentagon, the tanks used in the US invasion of Iraq had been upgraded in many respects. However, the enemy that awaited them would prove far more elusive than the Republican Guard of 1991, and undercut many of the army's assumptions about warfare.

Read the accounts of the Coalition's advance into Iraq in March 2003 and it is a tale of unequal slaughter, the proverbial 'shock and awe'. Crews fled their blazing tanks as Bradleys engaged them, the British knocked out a long line of armour leaving Basra, and M1s of the 3rd Infantry Division launched 'thunder runs' around Baghdad's ring road as the offensive drew to a close. If anything, the inequality in training and technology seemed even more pronounced than it had 12 years earlier.

But in the subsequent months and years, as insurgency gripped the country, things started to look very different. Faced with what the Israelis termed a 'disappearing enemy', every Coalition patrol or armoured vehicle could be attacked from any angle at any moment. Heavy armour was portrayed by the insurgents as a symbol both of oppression and the occupiers' fear of fighting on equal terms. As in Lebanon or Gaza, the destruction of tanks became a propaganda win, exciting Iraqis who shared images of blazing wrecks on their mobile phones.

For US soldiers or Marines, their missions would have been impossible without the protection of armoured Humvees or later the 'MRAPs' or Mine-Resistant Ambush Protected vehicles designed to survive typical insurgent attacks. And in the highest-risk areas they were very grateful for the support of M1 tanks, as I can attest from embedding on operations in which they were committed.

But whereas the Cold War had seen a technological contest, a battle over the means of attack and defence, requiring the development of advanced technologies and lasting decades, in Iraq the picture could change by the month. In insurgent or 'asymmetric' warfare, people could turn all manner of available technology to their advantage – like using a mobile phone to detonate an improvised explosive device. The British and Israelis had long experience of this cat-and-mouse battle of measure and countermeasure. So naturally, given Allied support and the imperative of not losing so many troops, the Americans stepped up their game quickly. Troop-carrying vehicles were transformed fast, but the tank was harder to adapt to this new environment.

The daily SIGACTS logs (logging significant acts of violence) put out by Multi-National Force Iraq charted the toll of bomb blasts and wrecked lives. And among those storied places, from Fallujah to Ramadi, Rustamiyah to Sadr City, they marked also the losses of tanks as operations progressed. The grim ticker recorded: the Abrams blown apart by a massive IED in Baqubah on the 23rd of October 2003, with the loss of two of its crew; one brewed up by multiple RPG strikes in Baghdad on the 18th of August 2004, four wounded; or the one knocked out by an Explosively Formed Projectile (or EFP) on Christmas Day 2005, with one crewman killed and two wounded.

This last incident was an early example of a shaped charge being used to defeat Coalition armour, a technology that Iranian Revolutionary Guards introduced to Iraq, sharing it with Shia and Sunni extremists alike as they sought to raise the blood price of America's occupation as high as possible. Thus the know-how of a state actor enhanced the performance of the insurgents' EFP bombs, and their knowledge of the Abrams' weak points.

It would be pointless here to recite the entire litany of SIGACTS involving US armour, but the salient point is that the insurgency, revved up by Iran, proved markedly more effective at destroying Abrams, albeit over a much longer period, than Saddam's Republican Guard had been in 1991. At that time, recall that 23 of the tanks were lost, but mostly to friendly fire, with just a handful, fewer than ten, destroyed by the Iraqis.

During the insurgency, reliable estimates suggest the number of M1s put out of action may have been ten times that number, around 80. The difference between a knocked-out tank and a destroyed one (i.e. whether it is repairable) often leads to lower numbers being used, but certainly one amateur researcher has logged 47 specific examples of these tanks being taken out of action.

Tank designers, from the Second World War onwards, had understood that the vehicle cannot be strong in all aspects, particularly if it is to have armour thick enough to defeat shaped charge (or HEAT) warheads. So they had assumed that when doing battle against another army, it would most likely be facing them and should therefore be most strongly protected at the front. Insurgents and more sophisticated non-state groups like Hezbollah have therefore learned to strike from underneath, from the flanks or rear, and increasingly from the top.

The Israelis spent 20 years prior to their assault on Hamas in Gaza reconfiguring their armour to take account of this. They deployed 'super heavy' armoured personnel carriers, and by the time the Mark 4 Merkava variant was fielded, it had a composite armour array on its turret roof that was 300–400mm thick. There was no such programme for the M1, which for example had as little as 25mm of armour on the turret roof. The lack of US investment in developing their fighting vehicles meant a heavier price was paid during the occupation of Iraq.

Different forms of attack could be applied to vehicle weak points by the insurgents. Massive IEDs, set off underneath a vehicle, could cave in its floor or even flip the whole thing over. When it came to the flanks or rear, shaped charges, often multiple ones called 'EFP arrays', could be set to detonate as a vehicle passed along the road. And with top attack, some very potent missiles like the Russian-made Kornet were available, but drones were increasingly adapted for this purpose.

In Iraq, the US developed up-armouring known by the acronym Tusk that was designed to give better (but not complete) protection against these threats. A set designed to cover the sides and rear against shaped charge attacks added 3 tons to the vehicle's weight, with another 2 tons being needed to protect the belly. For the M1 to have been as well protected in all aspects as it was frontally could have taken its weight up to the 90–100-ton bracket. Evidently this was not practicable for many reasons. With the Pentagon having scrapped its Future Combat Systems plan for a new generation of armour in 2009, relying simply instead on further M1 upgrades, once again the tank's future was called into question.

The war in Ukraine exposed more than a decade of under-investment in tanks, particularly by Nato countries. The same was true of everything from torpedoes to anti-aircraft missiles, the types of weapons used to fight the forces of other states rather than insurgent bands in Afghanistan or Iraq. Ukraine wanted tanks, but the M1 was the best and only option the US had available to send them. But with so many other sophisticated, costly weapons also in short supply, the task of rearmament is truly daunting. So does the tank have a future or not?

Extinction or Renaissance

Just prior to February 2022, and Russia's full-scale invasion of Ukraine, the tank appeared to be nearing extinction. While sceptics have questioned its vulnerability time and again from Cambrai onwards, the issue of sophistication and therefore affordability has been just as relevant.

A modern tank in 2024 is akin to a naval battlecruiser of yesteryear, complete with sensors, voice communications, datalinks and self-defence system. Some analysts talk of 'structural disarmament', a growth in sophistication to the point that each generation of weapons is so much more expensive than the last that numbers dwindle, leading eventually to extinction. In 2020, for example, the US Marine Corps, which had seen plenty of action in Iraq and Afghanistan, decided that the negatives, particularly relating to the M1's weight and size (given the expeditionary nature of their mission), outweighed the positives and that it should dispose of its tank battalions.

Speaking at the disbandment parade of his unit at Twentynine Palms, California, CO Lieutenant Colonel Benjamin Adams couldn't help reminding his audience that the Marines had hurriedly formed tank battalions at the outset of the Second World War, realizing they needed them, and 'my solemn hope is that, should the need for armour ever arise again, that spirit that has guided this battalion for the better part of a century, although asleep, will rise again.'

Adams's superiors had decided, though, that the money spent on such units would be better committed to 'sunrise' capabilities. Since they couldn't afford everything, they wanted the systems that were dawning, like drones and multiple rocket launchers, not the 'sunset' ones built for wars of the past. Similar discussions were happening across the Western Alliance.

In 2021 the UK's then Prime Minister, Boris Johnson, schooled the chair of the parliamentary defence committee, telling him, 'we have to recognize that the old concepts of fighting big tank battles on the European land mass are over.' Like the US Marines, Johnson made the argument that limited funds would be better spent on emerging technologies.

Prior to Russia's invasion of Ukraine, tank manufacturing was on the brink of extinction. Its global production figures had fallen to a post-war low, a decades-long trend but one that accelerated in the early 1990s after the end of the Cold War. In the US, the Lima facility was making – or rather rebuilding from reserve stocks – 11 vehicles each month or 132 annually. In 2021 Russia assembled just 70 T-90s at the Nizhny Tagil plant, with other historic centres of production in Omsk and Saint Petersburg reduced to rebuilding older models. Israel's production of the Merkava Mark 4 staggered on at about 50 a year.

This running-down of manufacturing simply reflected a cratering of demand. The Netherlands, whose army had 900 tanks at the end of the Cold War, followed the same principle as the US Marines and got rid of them altogether. In the UK the vehicle survived, but at a token level, the Challenger 3 project taking 150 old vehicles and upgrading them with a new turret. At the same time, the British finally dropped their resistance to the German 120mm smoothbore gun widely used across the rest of Nato.

These trends were given further impetus by the 2020 Azerbaijan and Armenia war. Dozens of Armenian armoured vehicles fell victim to missiles launched from Turkish-made Bayraktar drones as well as more basic unmanned aircraft using commercially available components. The vulnerability of armour to top attack was emphasized once more. Russian tanks, with their protection very much focused in the frontal arc, and their turret floor ammunition storage arrangements, proved particularly vulnerable.

Once hundreds of Russian tanks rolled into Ukraine in 2022

these weaknesses were exploited anew. Rudimentary attempts to protect their roofs from top attack with overhead metalwork, so-called 'cope cages', didn't seem to make much difference. Videos of Russian T-72s or T-80s 'turret flipping' after catastrophic internal explosions were widely shared on social media. Losses stacked up, so the Kremlin started to run down vast stocks of armour it had parked in storage sites after the end of the Cold War.

Nevertheless, the Ukrainian military leadership appealed for supplies of Western-made tanks. And indeed, when it came to launching its 2023 counter-offensive, as noted at the start, President Volodymyr Zelenskyy argued that their supply was key to liberating his land from its Russian occupiers. What he got in response to these appeals was a motley collection of Western types – from the modern Leopard 2s or Challenger 2s to older Leopards and rebuilt Soviet models. An American commitment to send 31 M1s took a little longer to materialize.

The performance of the American machines, when they finally did appear on the battlefield early in 2024, just seemed to underline the challenges faced by all designs conceived decades before for head-to-head fighting. Straight off there were grumbles about the hefty M1 sinking in the mud. In eastern Ukraine, the old criticism of the M1 being an 'autobahn tank', because of its weight, had real meaning. And the labour of keeping its engine air filters clean, one all too familiar to American tankers, also irked its Ukrainian crews.

Assailed by drones (including cheap First Person View or FPV ones guided to their targets by operators using virtual reality goggles or even smartphones), the weakness of the M1's side, rear and top armour was exploited, causing the rapid loss of one-third of the American tanks supplied and their temporary withdrawal from service.

In this conflict the Bradley, with its ability to deliver infantry assaults as well as stand its ground against other armour with a 25mm gun and anti-tank missiles, has proven very popular with Ukrainian crews. While it also has composite armour, its weight is half that of an M1 Abrams.

Might this show the way of the future: a well-armoured, tracked vehicle with lots of different uses, but at a more manageable weight? It may do, though the fact remains that in a one-to-one fight a tank is more likely to prevail over a Bradley because of its big gun and heavier protection.

Ukrainian M1 crews, in common with those sent to fight in these machines across the decades, knew that a tank could become a magnet for fire for all kinds of enemy weaponry. Along with changing their tactical approach, they also fitted their M1s with bolt-on armour to cover their weak points. In desperation at their own losses, the Russians deployed what was nicknamed the 'turtle tank'. It was fitted with such a big framework of protective cages, designed to block drones from all angles, that it looked like a house on the move.

Yet despite all this evidence that tanks have become expensive sitting ducks, armies have not given up on them. Far from it. The Poles ordered 1,000 K2 tanks from South Korea soon after the invasion of Ukraine. Faced with the ever-present threat of invasion from the north, the South Koreans, unlike most Nato countries, had maintained an extensive defence industry, which reaped the rewards following Russia's invasion of Ukraine. Romania in 2023 announced a $2.5 billion order for 51 (rebuilt) M1s. That contract, including spares, training and support, worked out at $46 million per vehicle! So much for the $1m psychological barrier that appalled Congress members several decades earlier. Russia has attempted to increase tank production to 300 a year. Israel, embroiled since October 2023 in its war against the Palestinian Hamas movement, has greenlit the production of the Mark 5 Merkava, a major investment.

To understand why, we need to return to that simplest of ideas, the one that seized the British generals facing stalemate in 1916. If you can find a way to move powerful weapons around a battlefield criss-crossed by fire under armoured protection you will want to exploit it – if only to reduce the casualties being suffered.

'What saves tanks, time after time, is war,' remarks Israeli tankist Major General Guy Tzur. When Major General Kathryn Toohey concluded that 'tanks are like dinner jackets. You don't need them very often, but when you do, nothing else will do', it was in the context of a big Australian study into what weapons it required for the future.

So what are those needs? It's important to distinguish between armoured vehicles in general and tanks specifically. If you want to move troops across a battlefield swept by small arms and artillery fire, you will need to protect them. That's an imperative as obvious today as it was to British generals on the Somme in 1916. Now as then, there are wheeled vehicles that can be used to do that, often quite cheaply. Once the list of things you're trying to protect them from extends to weapons heavier than rifles or hand grenades, armour gets thicker, so the vehicle's weight starts to increase exponentially.

Heavily armoured vehicles need tracks to stop them sinking into the ground as soon as they leave metalled roads. To the layman, any armoured vehicle on tracks, particularly one with a turret, will be called 'a tank' even if its primary purpose is carrying infantry or a heavy artillery piece. However, the soldier knows that nothing will kill those various kinds of armoured vehicles more effectively than what *they* call a tank. Why? Because this type of vehicle is armed with a gun capable of imparting to a penetrating dart or shot so much kinetic energy that it can punch through any target it's likely to encounter. There's something reassuringly low-tech about that.

The latest automated protection systems may be able to shoot down an anti-tank missile. Explosive reactive or composite armours may neutralize its shaped charge warhead. But the tungsten, steel or depleted uranium penetrator, flying at 1,800 metres per second, is far harder to stop. So if they are to remain on the battlefield, the surest way of killing one tank is with another.

Add to that the ability of a tank to fire many other types of ammunition against various targets, responding instantly to the needs of the troops it accompanies, and perhaps the understanding of why sometimes nothing else will do becomes clearer. Tanks continue to provide high levels of protection to powerful, mobile guns aimed through optics that far exceed the capabilities of the human eye.

Night vision or other advanced optics are features of the recent past. But this triad – protection, mobility and firepower – has been a factor throughout the life of the tank. History has repeated itself many times, for example in the discovery during development or combat that the armour planned is inadequate. Over time this has meant a *fortyfold* increase in frontal protection: from 14mm on Mark IVs to 30mm on early Panzer IVs, the Tiger's 100mm, 400mm on the T-64's turret, and eventually 600mm on some M1s.

Engine horsepower has gone up, but not by nearly as much – it's a fourteenfold increase between the Mark IV and M1. So while the modern vehicle might harness its horsepower more efficiently through a better transmission, suspension and running gear, we can see that tanks have in relative terms become larger and more ponderous over time. General Eisenhower's prediction of the vehicle ending up as a steel pillbox seems to have been right.

We've also seen, across the century-plus of this story, a steady growth in the complexity of these machines. A Renault FT, with perhaps 1,500 component parts, evolved into a Sherman with 4,300, a Centurion with 30,000 and an M1 or Merkava with many times that. This transformation naturally saw costs balloon too. The British tank of 1917 cost £5,000 or around £291,000 in 2024 prices, the new Challenger 3 around £5 million per vehicle. This is why cost, along with weight and complexity, has also had its effect, contributing to the shrinking of tank fleets worldwide.

They continue to be useful to armies, though. In current conflicts an ongoing need for high protection levels has caused commanders to employ them in places where they would have hesitated in the Second World War. For the Russians in Mariupol or the Israelis in Gaza, the tank forced its way into urban areas. Without them (and personnel carriers with similar protection levels in the Israeli case), their losses would just be prohibitive.

If the Ukrainian army received some daunting lessons when

trying to break through with their Leopard 2s or M1s, they simply served to underscore Guderian's or Montgomery's ideas about the need for combined arms warfare and control of the air. Even so, the apparent stasis along much of the Ukrainian front suggests that a turn of the technological wheel is needed to restore some mobility.

In order to get moving, you still need to survive. And on the current battlefield the competition between means of attack and defence has accelerated. Disruptors, from Middle Eastern insurgents to Ukrainian tech bros, have evolved new ways of attacking weak spots. Now we see the spread of electronic jamming systems to spook drones, and protection systems to shoot down missiles as they fly towards the tank. New generations of armour are under development too, as well as experiments with different vehicle layouts. By taking soldiers out of the turret, for example, its size can be reduced, and the weight saved given to thicker protection.

New materials are constantly being developed also to improve the mass efficiency of armour, keeping the vehicle's weight (just about) manageable. We've seen how reducing internal volume can also help with this. The eight-man crew of the Mark IV in 1917 gave way to the five-person one, typically, in the Second World War, with that going down to four or even three during the Cold War. Ultimately, doesn't that logic suggest we will end up with robot tanks?

Early experiments with 'unmanned ground vehicles' have yet to yield a breakthrough. Bereft of a human driver or commander, they are more likely to get stuck, less so to contribute effectively on a dynamic battlefield. But ultimately a crewless armoured fighting vehicle appears inevitable.

Seeing the thousands of knocked-out tanks on the Ukrainian battlefield, crematoria for so many soldiers, the possibility that people might be spared the experience of going to war in these machines must surely appeal. Just as the British invested in the armour of 1916 in the hope that it would save their soldiers' lives, so the robotic vehicle whispers that promise of the First World War tanks once again: of great victories bought at a lesser price in blood.

Acknowledgements

This journey began as a conversation between me and Daniel Crewe, the editor at Penguin with whom I wrote my last book. We are it seems surrounded by war these days, with participants and onlookers alike constantly pointing out the vulnerability of the tank one day and its indispensability the next. Oddly, this argument has been around since the first armoured vehicles chugged into action in 1916 during the Battle of the Somme and is heard again with each major conflict.

It is a machine that symbolises the brute horror of war in the industrial age. Yet at the same time, its armoured shell is essential to saving life, from its origins on the bullet-swept quagmire of the First World War to eastern Ukraine today.

While attitudes to it have fluctuated in these familiar ways, the tank itself has evolved to a remarkable extent. The story of its development, manufacture, and use could easily fill many volumes, which is why histories of the tank per se are rare things. It is also why, even when telling the story through ten examples, a very broad sweep of historical research is required.

I am very grateful therefore to Thomas Anderson, Tim Gale, Colonel Benny Michelsohn, Stephen Walton, Wen Jian Chung and Steve Zaloga for their expert guidance in my research. Particular thanks are due to David Willey and Marjolijn Verbrugge, the curator and librarian respectively at The Tank Museum in Bovington, for all their help during my research trips there.

Editing a book like this is never a task for the faint-hearted so I'm very thankful for the tender mercies of Daniel Crewe, Alexandra Mulholland and Duncan Heath, whipping my jottings into something more coherent. Once again, as he has peerlessly for decades, my agent Jonathan Lloyd did the deal on my behalf.

Lastly my home hero, my wife Hilary, who keeps the love and support advancing inexorably, crushing doubts and sustaining my literary advance. I could not be better backed.

Any mistakes in these pages are of course entirely down to me.

Index

Page references in *italics* indicate images.